The Universe

The Universe

It's not what you think!

By

Keith Dixon-Roche

(with a little help from Isaac Newton)

The Universe

Keith Dixon-Roche © 2019 to 2026

All concepts and formulas in this book not previously attributed to 'The Heroes' identified in The Appendix, are the sole property of Keith Dixon-Roche and protected by copyright.

Their use, publication, broadcasting, distribution, copying or any form of recording without Keith Dixon-Roche's written consent shall be a breach of international copyright law and subject to legal action.

The Universe

It's not what you think!

Published by CalQlata

info@CalQlata.com

First published November 2019
Final publication April 2026

This book is sold subject to condition
that it shall not by way of trade or otherwise,
be lent, re-sold, hired out or otherwise circulated
without the publisher's prior consent and in such
circumstances it shall not be circulated in any form of
binding or cover other than that in which it is published

Copyright © Keith Dixon-Roche 2019 to 2026

Contents

Preface		9
1	Introduction	11
2	Quanta	13
2.1	The *'Photon'*	14
3	Proton-Electron Pairs	17
4	Atoms	19
4.1	Elements	20
4.2	Nucleus	21
4.3	Electron Shells	22
4.4	Isotope	23
4.5	Ion	24
4.6	Fission (Radioactivity)	25
4.7	Fusion	27
4.8	Molecules	28
5	The State of Matter	29
6	Energy	31
6.1	Potential	34
6.2	Kinetic	35
6.3	Electrical & Magnetic	36
6.4	Electro-Magnetism	37
6.4.1	Wave Characteristics	39
6.4.2	Heat & Light	41
6.5	$E=mc^2$	44
7	The Universal Machine	47
7.1	Orbits	48
7.1.2	Terminology	50
7.1.3	Orbital Laws	52
7.1.4	Station-Keeping	55

7.1.5		Orbital Planes	60
7.1.6		The Importance of Orbits	61
7.2	Spin		62
7.2.1		Polar Moment of Inertia	65
7.2.2		Earth's Core	66
7.2.3		Earth's Magnetic Field	67
7.2.4		Magnetic Reversal	68
7.2.5		No Moon!	69
7.2.6		Chicken & Egg	70
7.3	Core-Pressure		71
7.3.1		The Structure of Celestial Bodies	72

8 A Universal Theory (how it works) — 73

8.1	The Ultimate Body	76
8.2	The Universe	77
8.2.1	Universal Size and Age	78
8.3	Planetary Spin Energy	79
8.4	Galaxies	80
8.5	Solar Systems	81
8.6	Stars	82
8.7	Planets	83
8.8	Moons	84
8.9	Our Galaxy: The Milky Way	85
8.9.1	Hades	86
8.9.2	Our Sun	87
8.9.3	Mercury	88
8.9.4	Venus	89
8.9.5	The Earth	90
8.9.6	Mars	92
8.9.7	The Asteroid Belt	93
8.9.8	Jupiter	94

8.9.9	Saturn	95
8.9.10	Uranus	96
8.9.11	Neptune	97
8.9.12	Pluto	98
8.9.13	Our Moon	99
8.9.14	Phobos	99
8.9.15	Deimos	99

9   Fact & Fiction   101

9.1	Sub-Atomic Particles (fiction)	102
9.2	Black Holes (fiction)	103
9.3	Big-Bang (fact)	104
9.4	Dark Matter (fiction)	106
9.5	The Birth of Our Solar System (fiction)	107

10   Primary Constants   109

10.1	Electrical Charge (e)	110
10.2	Magnetic Charge (kg)	111
10.3	Distance (R_n)	113
10.4	Time (t_n)	114
10.5	Static Ratio (ξ_m)	115
10.6	Dynamic Ratio (ξ_v)	116
10.7	Universal Constant (Σ)	117
10.8	Temperature (T_n)	119

11   Model Verification   121

11.1	Density vs Temperature	122
11.2	Specific Heat Capacity	123
11.3	Gas-Point	124
11.4	Our Sun	125
11.5	PVRT	126

12   The Laws of Thermodynamics   127

13   So; What Now?   129

Appendix		131
A1	References	133
A2	Glossary	135
A3	The Heroes	137

The Universe

The Universe

Preface

I am aware of two genuine scientific discoveries in the 20th century, that were not made as a result of guesswork; DNA and continental drift, neither of which were made by the scientific community.
Rosalind Franklin discovered DNA, she died making it. Her boss - a member of the scientific community - stole her work and gave it to a couple of his scientist friends. She was ignored by the scientific community who attributed her work to the friends; the names of whom I have forgotten.
Alfred Wegener discovered continental drift. The scientists of his day disagreed with him, calling him; "an amateur"; "a jumped-up meteorologist"; etc., however, as we now know, he was right and the scientists were wrong. To my knowledge, not one member of the scientific community has had the humility to publicly apologise for their disgraceful behaviour.

Moreover; I was involved in the **JET project** back in the 1970's. When I discovered that the project was based upon stellar fusion, I told one of the scientists that it could never work because the sun's energy was from fission, not fusion. The scientist told me to mind my own business. Here we are 50 years later and I have now shown that stellar energy does indeed come from fission and we still don't have a fusion reactor.

The atom bomb was also created by experimentation and misunderstanding. Nuclear energy is still misunderstood today, if it were not, we would not now be stuck with Chernobyl, three-mile island and the storage of nuclear waste. If the scientists understood it, they would also know that nuclear energy can be extracted from any matter, and it can be switched on and off at will; safely, cleanly and efficiently, and its generators cells can be any size.

Following my recent studies of Newton's laws of orbital motion, along with the discoveries I have made using them, I have come to the conclusion that the scientific community made a few spectacular errors back in the early 20th century that it continues to perpetuate, simply because it is afraid to admit that it may be wrong.

Because I have never been convinced of the models of the universe I was taught by my scientific tutors, I recently decided to test their theories using mathematics and verification; and you guessed it, I discovered they were wrong, again.

Our universe is very different to the one we have been taught.

The Universe

Because it is now possible to define the Milky-Way's force-centre, I have given it the name 'Hades' for easier reference.

Keith Dixon-Roche 2026

1 Introduction

This book is a non-mathematical overview of the things that make up our universe, which comprises more than 2.8E+75 protons together with the same quantity of electrons, and nothing else. I shall begin my overview with these particles (Chapter 2); because there is nothing else out there.

Max Planck called these particles; **Quanta**, which in deference to him, being the only 20th century scientist needed to describe our universe, shall be used as a collective term for all *particles* in this publication.

Mass, as we understand it is actually 'magnetic charge'. It is not just *possible*, it is actually very easy to calculate the mass of a body, such as a planet, using magnetic charge#. All physical properties can be calculated from just four basic constants (distance, time and electrical & magnetic charge), two ratios (static & dynamic) and a particle property of 3E-91#

The universal model described here is very different to the one you know. It is based upon scientific and mathematical theories generated by the heroes listed in the Appendix of this book, almost all of which originated well before the 20th century.

Every concept presented here, from the atom to the '*Big-Bang*', is verifiable mathematically and physically, and they all interrelate. At no time has it been necessary to qualify any claim with the term "*I think*".

This is a mathematical universe that needs no unification theory or sub-theory to justify it. It obeys; Newton's conservation of momentum, the laws of thermodynamics and the conservation of energy. It is eternal, and needs no '*outside*' assistance.

Whilst mass is actually magnetic charge and gravity is magnetism#, they will be referred to as 'mass' and 'potential energy' respectively in this book in order to prevent confusion for the reader.

It is important to understand that *all* natural laws are simple and *locked*; i.e. there can be no possibility of change. Statistics apply *only* to the consequences of laws, never the laws themselves.

Whilst this book includes the principal mathematical proofs, additional references may be found in the publications listed in the Appendices.

refer to my earlier publication 'The Physical Constants'

The Universe

2 Quanta

There are only two Quanta in the universe; the proton and the electron. Every proton is identical to every other proton, and every electron is identical to every other electron.

Electrons are dynamic and possess negative electrical and non-polar magnetic charge.

Protons are static and possess positive electrical and non-polar magnetic charge.

Whilst protons and electrons have the same density (7.1266E+16 kg/m³), the proton is 1836.15 times more *massive* than the electron.

The magnetic charge in all Quanta is constant; it never changes.

Fig 1

The electrical charge in electrons is constant; it never changes.

The electrical charge in protons varies with the velocity of its orbiting electron.

Neutrons are proton-electron pairs united through high-temperature due to planetary spin in bright stars. All of the neutrons in all universal elements were created in previous universal periods.

Contrary to all you may have been taught, these Quanta are discrete, 'stand-alone' packets of electro-magnetic charge. They are not composed of sub-atomic particles (quarks, fermions, leptons, gluons, etc.).

The Newton-Coulomb atom works perfectly well without them.
Why would nature divide perfect particles into sub-particles that compromise reliability and waste energy? If there is one thing certain about nature, it does not waste energy on things it doesn't need.

Nature's Quanta are simple and perfect; there is nothing in them that can go wrong, and they always do exactly what is expected of them.

2.1 The *'Photon'*

Photons are a figment of our imagination.

They are said to exist because scientists believed that Crooke had created a perfect vacuum in his tube, and that the light emanating from within must therefore be electrons travelling at the *'speed of light'* (photons). Crooke's tube was, however, anything but empty; it still contained at least 1E+12 protons. The light seen was actually electro-magnetic energy generated by interaction between the stationary protons and the travelling electrons inside the tube.

Such an interpretation of light is difficult to understand given that; if the wavelength of the entire electro-magnetic spectrum ranges between 1.77E-14m and 0.0943m, how can all photons travel at the same velocity. Surely, they must travel at all speeds between 299792459m/s and 17162.242521927m/s in order to represent the full electro-magnetic spectrum.

For example; the kinetic energy of an electron travelling at 1E+06m/s will be different to another travelling at 1E+08m/s. Therefore, electro-magnetic energy (e.g. colour) that each radiates must also differ.
And if light possesses mass, different wavelengths must be deflected at different angles according to Newton's laws of gravitational attraction. Which doesn't happen: *All wavelengths of light passing the same celestial body at the same radial distance are deflected at the same angle.* Note: rainbows occur due to wavelength variation, not gravitational force. Therefore, light cannot possess mass, Einstein should not have used Gravitational force to calculate deflection.

Moreover, special relativity was devised because of the inability to correlate the additive nature of mass-velocity with the non-additive nature of light. This is only a dilemma if light possesses mass, which it doesn't; confirming this misunderstanding of the nature of light.

As a result of this error, all physicists now claim that the light we see is emitted by photons. They also claim that electrons are 'weird beasts' that possess mass and travel in waves, which is the reason we cannot pin them down (uncertainty principle).

It was this misunderstanding - that light is electrons and therefore possess mass - that caused Einstein and Bohr to propose their theories.

Not only is it unnecessary to deform space-time around celestial bodies in order to explain light deflection, it is mathematically incorrect to do so.

The light we see cannot be photons. It must be electro-magnetic energy, which possesses no mass and is deflected by magnetic charge.

The light radiated by the stars is simply our brain's interpretation of electro-magnetic energy emitted by proton-electron pairs at wavelengths between; 8E-07m & 4E-07m (4,900 K & 7,800 K), all of which travel at the same velocity; 299792459 m/s

If light is electrons, our sun would have emitted all of its electrons within the first few seconds of its life leaving behind a cloud of lone protons, which of course, was not the case.

If only Einstein, Bohr, Thomson, etc. had realised that Crooke couldn't possibly have created a perfect vacuum in his tube, scientific and technological development would be 100 years further ahead than it is today.

There are no such things as photons.

The Universe

3 Proton-Electron Pairs

The fundamental universal partnership comprises one proton orbited by a single electron. It is called a proton-electron pair. This partnership is made possible by the opposite polarity of their electrical charges.

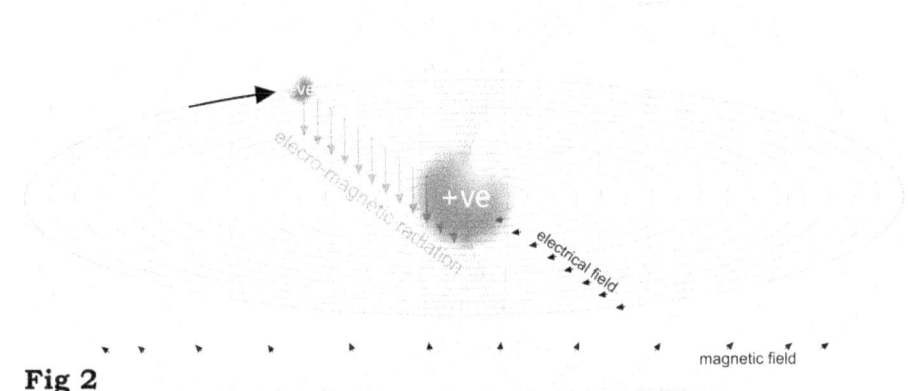

Fig 2

Orbiting electrons collect electro-magnetic energy from their environment, convert it into kinetic energy (velocity), and simultaneously transfer it to their proton partners via their opposite electrical charges. The proton partner uses its additional magnetic charge (x 1836.15) to collect this energy, which the pair then emits as electro-magnetic radiation. This cycle continues until the electron is physically removed from its orbit or possesses insufficient energy to maintain it.

Electron orbital radius varies with kinetic energy. The velocity of an orbiting electron increases with increasing electro-magnetic energy, causing its orbital radius to decrease.

If the orbiting electron collects sufficient electro-magnetic energy to achieve the '*speed of light*', the two particles will unite to create a neutron. This phenomenon occurs within bright stars and releases fissionable energy.

Atomic proton-electron pairs radiate electro-magnetic energy according to their elemental lattice structure, and are therefore responsible for an element's distinctive Balmer lines.

Electrons can only be removed from their proton partners either by force (impact from another electron) or electricity (voltage – potential energy).

All electron orbits are circular, where 'PE = 2.KE', and therefore the essence of Henri Poincaré's formula;

$E = PE = 2.KE = 2 \cdot \tfrac{1}{2}.m.v^2 = m.v^2$

Important Note:
'E' in Poincaré's formula has nothing to do with kinetic energy; it defines potential energy.

It is not known *how* an electron and its proton unite as a neutron. It may be, that the proton engulfs its electron partner, but we do know that once united, this union has a limited natural life (half-life). It will separate into its component parts; a proton (alpha particle) and an electron (beta particle) again after a given period, and we also know that this life is dependent upon the neutron-neutron interaction within an atomic nucleus.

4 Atoms

Atoms are collections of proton-electron pairs. They are created under very high pressure whereby the proton of one proton-electron pair is forced inside the electron shells of another. This is only possible, however, in atoms that have acquired neutrons during previous universal periods and within a body of Quanta sufficiently cold and massive to overcome the repulsion in adjacent elemental protons, due to core pressure. The ultimate body, the great attractor and galactic force-centres are the only masses capable of such forces naturally.

It is a fundamental law of nature that a satellite can have only one force-centre, therefore, even within an atom each electron will remain tied to its original proton partner, irrespective of its energy or configuration (Fig 3).

When an atom achieves a neutronic ratio (ψ) greater than 1.5 by increasing its neutron population (in bright celestial bodies), it will readily and regularly eject neutrons as alpha (protons) and beta (electrons) particles thereby releasing their energy (nuclear fission).

Fig 3

Atoms are identified by their atomic number (Z), i.e. the number of protons in their nucleus. The natural range of atoms is $1 \leq Z \leq 92$.

4.1　Elements

An element is an atom including its neutrons, that are responsible for an element's characteristics; half-life, lattice-structure, mass, chemistry, etc.

There are three types of hydrogen atom all of which radiate EME, but only two of which are the building blocks of elements;

    Hydrogen; the proton-electron pair

    Deuterium; an hydrogen atom with one neutron attached

    Tritium; an hydrogen atom with two neutrons attached

All elements comprise collections of deuterium and tritium atoms.

This means that an element's neutronic ratio (ψ) could, in theory, be between 1 and 2, however, this is not possible because of the neutron-neutron interaction. I.e.

    $1.66666 < \psi \leq 2.0$　　impossible

    $1.6 < \psi \leq 1.66666$　　instant self-destruction

    $1.5 \leq \psi \leq 1.6$　　radioactive

    $1.0 < \psi < 1.5$　　normal

    $\Psi = 1.0$　　chemically inert

All elements are continually trying to lose their excess neutrons ($\psi > 1$), i.e. they're continually trying to reduce their tritium atoms to deuterium atoms. The rate at which they do this is called their half-life. If they actually achieve it, the element will become chemically inert. Luckily however, the time over which this occurs in normal elements is many times longer than a universal period (31.644 billion years)

Elements are identified by their names; hydrogen to uranium.

Technetium ($Z = 43$) is a special case; it is highly radioactive despite having a neutronic ratio of just 1.300144.

Neutrons are only created inside atoms and cannot exist outside them.

The Universe

4.2 Nucleus

The atomic nucleus is a collection of protons and neutrons arranged in a lattice structure that minimises proton-proton and neutron-neutron interaction. The electrical repulsion pattern is therefore responsible for creating crystals in same-element atoms in both viscous and gaseous states.

For instance; what we currently refer to as; body-centre cubic, face-centre cubic, close-packed hexagonal, tetrahedral, etc. defines the structural arrangement of the protons in an element's nucleus. Which in turn defines the electrical charge arrangement between adjacent elements in viscous matter, and is also responsible for the even-distribution of adjacent atoms in a gas; Dalton's theory (refer to Chapter 11.5).

These lattice structures are responsible for the lattice structures in elemental matter. Nucleic protons (and their neutron partners) do not touch each other; they space themselves naturally to minimise neutron-neutron and proton-proton interaction. Moreover, elemental electrons do not orbit a nucleus, they orbit their proton partners.

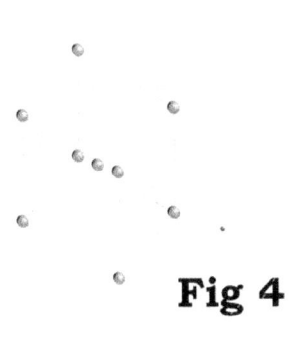

Fig 4

Each proton in an atomic nucleus is part of a proton-electron pair. It always was and will remain as such until its electron is physically removed. Following such an event, e.g. when neighbouring electrons swap atoms during DC electricity, all outer electrons instantly and automatically redistribute to fill the gap and balance their electrical charges.

4.3 Electron Shells

There are only two orbital electrons per shell radius in every atom.

The orbital path of each atomic electron is circular, about its proton partner. Therefore, whilst there are always only two shells of the same radius, each shell will be offset according to their proton's nucleic arrangement.

The spacing between shells is identical to the orbital radius of the innermost electron shell. And the orbital radius of each pair of shells decreases with increasing velocity (energy).

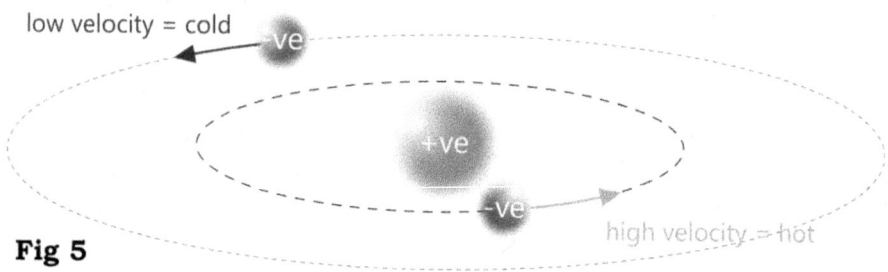

Fig 5

In viscous matter, shell radii are always much larger than the distance between adjacent atoms (d = $\sqrt[3]{[m_a/\rho]}$), which results in electron clouding and therefore molecular and atomic bonding. The higher the heat energy in the matter, the smaller the shell radii and the lower the electron clouding; reducing molecular and atomic bonding. There is no electron clouding in gaseous atoms.

There are no shell valences. When an electron is removed from an atom's shell structure, all the outer electrons will relocate to fill the gap. This is essential to ensure that all electrical forces balance.

4.4 Isotope

Isotopes are atoms with the same atomic number (Z) but with varying atomic mass because of unequal proton-neutron pairing. Isotope is an alternative way of saying RAM (relative atomic mass).

An atom of iron, with 26 protons (Z=26) and 26 neutrons (N=26) is an isotope of 52. However, in nature, most iron atoms have more than 26 neutrons, each of which is given its own isotope, e.g. 57, 59, etc.

The following rules apply to isotopes:
1) H⁺ can never be fused because it only exists as a gas
2) All proton-electron pairs within atoms are Deuterium or Tritium
3) If $ψ$ = N:Z (ratio), then: $1 < ψ < 2$ (see below)

Despite the potential maximum value for $ψ$ = **2**:

*If an atom achieves a 'ψ' value of greater than **1.5** it will readily eject neutrons as alpha and beta-particles (radioactivity).*

*If an atom achieves a 'ψ' value of greater than **1.6** it will split into smaller atoms ejecting numerous alpha and beta-particles as it does so.*
***1.6** is the limiting number for isotopes.*

Over time, atoms naturally try to achieve $ψ$ = **1**, which is their most stable form. They eventually achieve this by ejecting surplus neutrons as alpha and beta-particles. The rate at which this occurs is referred to as the 'half-life' of the atom. The half-life of any atom appears to be constant, i.e. it never appears to change.

4.5 Ion

Ions are atoms with the same atomic number (Z) but possess an electrical charge owing to unequal proton-electron pairing.

Positive ions (atoms that have lost electrons) possess a positive electrical charge. Negative ions (atoms with additional electrons) possess a negative electrical charge. Negative ions are far less common than positive ions.

Only a few atoms exist naturally as negative ions and they are all non-metals$_n$ except for two, which are semi-metals$_s$:

One additional electron (Group VIIA):
Fluorine (9_n), Chlorine (17_n), Bromine (35_n), Iodine (53_n)

Two additional electrons (Group VIA):
Oxygen (8_n), Sulphur (16_n), Selenium (34_n), Tellurium (52_s)

Four additional electrons (Group IVA):
Carbon (6_n), Silicon (14_s).

Any atom can become a positive ion simply by losing one or more of its electrons from impact with free electrons or a strong external positive electrical charge.

Negatively charged ions are a little more difficult to understand. Additional electrons need to be trapped by the positive charge in protons that do not exist in the nucleus: this shouldn't be possible. However, the nucleic structures of the above non-metal atoms probably have at least one exposed proton that is not protected by a neutron and this means that the additional electro-magnetic electrical charge generated in it is available to trap passing free electrons

4.6 Fission (Radioactivity)

Fission is the splitting of neutrons into their component parts (a proton and an electron), releasing their stored energy.

The ejection of neutrons is called radioactive decay, and the time over which it occurs is referred to as its half-life, and it releases a great deal of:
electro-magnetic energy if the neutron decays into a proton-electron pair but is unable to escape from the atomic nucleus. In this case, the energy is released as heat and the atom concerned will have become a different element (Z+1); or,
kinetic energy, if the neutron is released from the atom. In this case, the proton and the electron will be ejected at a velocity of; $v = \sqrt{[2.m_p/E]}$ (7E+06 m/s), the proton of which will impact the neutrons in neighbouring atoms, splitting the impacted neutrons into their component parts (protons and electrons), the protons of which will then split other neutrons. This is called a chain reaction.
Radiation is normally referred to as the ejection of helium atoms (alpha-particles [2 protons] + beta-particles [2 electrons]), because:
Natural Fission: As neutrons are created in pairs, their half-life will cause them to be ejected in pairs; or.
Unnatural Fission: An ejected proton will never impact a neighbouring proton because of their similar electrical charges. Neutron loss through impact may expose neighbouring protons, resulting not only in the loss of a neutron but also the proton it was isolating.

Whilst neutrons are continually ejected by all atoms with a neutronic ratio greater than '1', we refer to those with a neutronic ratio greater than '1.5' as unstable because they represent a danger to life-kind.

Atoms close to their critical ratio are subject to a limiting mass condition above which the atomic matter will begin to break apart. This condition is called an atom's critical mass.

As the critical mass is approached, neutron targets in neighbouring atoms are increased resulting in a consequent increase in temperature. Or, if the condition is enforced quickly enough, the ejected atoms will have nowhere to go so the matter will break apart. This chain reaction is what occurs in an atom bomb.

The reason the aftermath of an hydrogen bomb is not radioactive is because its by-product is hydrogen. Whereas only a very small percentage of the matter in a uranium or plutonium bomb is converted to non-radioactive matter, the rest of it is scattered as radioactive dust, which is extremely dangerous to anybody coming into contact with it. However, the uranium used to initiate an hydrogen bomb also makes H-bombs radioactive.

The reason why much more energy is released in the detonation of an hydrogen bomb is because very little of the energy released is lost in splitting atoms, *they are already split.*

Whilst our knowledge today only allows us to extract neutron energy from radioactive matter, it can be released safely from any and all matter if processed correctly in a controlled manner.

Neutron energy can, and should, be regarded as a friend, not an enemy.

4.7   Fusion

Fusion is the union of proton-electron pairs and/or atoms to create a different element. It is accomplished by applying sufficient force to push the proton of one proton-electron pair inside the electron shells of an atom.

Atoms can only fuse if they are in viscous form. You cannot fuse gaseous atoms because the repulsive electrical charge energy is greater than the attractive magnetic field energy.

Atomic fusion can only occur inside bodies with sufficient mass to generate the necessary core pressure. This is only possible inside galactic force-centres and the ultimate body.

When atoms or proton-electron pairs are fused together, a small amount of electro-magnetic energy will be released as their electrons rearrange themselves into shells. The energy released, however, will be considerably less than that required to fuse the atoms. For example, @ 30°C:

The kinetic energy in a carbon atom = 7.29997E-20 J
The kinetic energy in an iron atom = 1.26627E-19 J
Individually, they generate a total of: 1.99626E-19 J
United (as Germanium), they would generate: 1.34614E-19 J
Releasing: 6.50124E-20 J

The potential energy in a carbon atom = -1.45999E-19 J
The potential energy in an iron atom = -2.53253E-19 J
The energy required to unite these two atoms would be: -3.8598E-19 J

Representing a nett energy input of: -3.20968E-19 J

Therefore, it is necessary to input 3.2E-19 Joules of potential energy in order to release 6.5E-20 Joules of kinetic energy; in other words, you need to input five times as much energy as you're releasing.

Fusion requires the *input* of energy; it does not generate energy. That is why Hades is cold and 40 years of trials have yet to produce a fusion reactor.

4.8 Molecules

Electrical field energy (EFE) bonds atoms together as molecules in the same way that MFE bonds atomic Quanta (refer to Chapter 11.1).

In theory, any two atoms can form an EFE bond, but most will be unstable. Stable atomic bonds tend to include small atoms with low 'Γ'# values, e.g.:

Hydrogen (Γ = **0.07146**)
Helium (Γ = 0.011709)
Carbon (Γ = **0.01605**)
Nitrogen (Γ = **0.0086143**)
Oxygen (Γ = **-0.0001125**)
Fluorine (Γ = 0.998403)
Neon (Γ = 0.16173)
Sodium (Γ = 0.8098118)

Magnesium (Γ = 0.22875)
Aluminium (Γ = 0.6795263)
Silicon (Γ = 0.0549643)
Phosphorus (Γ = 0.5842566)
Sulphur (Γ = 0.0365625)
Chlorine (Γ = 0.7692353)
Potassium (Γ = 0.5202474)
Calcium (Γ = 0.0351)

The smaller the atom and the lower its 'Γ' value the more stable its bond. Therefore, most molecules tend to include those above highlighted in **bold type**. Whilst sulphur, calcium & silicon all have low 'Γ' values, their size inhibits stability.

Helium is the odd-man-out here because whilst it is small and has a low 'Γ' value, it strongly resists molecular partnerships because it has no spare neutrons and only one electron shell, which is full; it is perfectly balanced.

The following natural bonds form between same-element atoms (diatomic) in gaseous form:
Hydrogen (H_2), **Nitrogen** (N_2), **Oxygen** (O_2), Fluorine (F_2), Chlorine (Cl_2), Bromine (Br_2), Iodine (I_2)
Size inhibits stable molecular bonding in Bromine & Iodine with anything but same-element atoms.

The lowest 'Γ' values can occasionally generate triple bonds ...
Carbon, Nitrogen, Oxygen (e.g. O_3)
... but they tend to be very unstable (e.g. ozone)

Because of the instability of molecules with high 'Γ' values it must be concluded that neutrons inhibit atomic bonding, which supports the premise for electron clouding (refer to Chapter 5)

an atom's 'Γ' value is a mathematical description of its nucleic, and therefore, its lattice structure; $Γ = 9.[ψ-1]$ where; 'ψ' is the atoms neutronic ratio (N:Z)

5 The State of Matter

There are only two forms of matter; viscous (solid and liquid) and gaseous.

Adjacent atoms are held together by the magnetic field energy generated by an atom's proton-electron pairs, and they are pushed apart by the electrical charge in their nucleic protons.

If the magnetic field energy is greater than the electrical charge, matter will exist in its viscous form. If the electrical charge is greater than the magnetic field energy, matter will exist in its gaseous form.

Because the magnetic charges are constant, the magnetic field they generate is also constant. Hence matter density does not alter [significantly] with changes in temperature in viscous form.

Because proton electrical charge varies with temperature, the forces pushing adjacent atoms apart also varies. Hence unconstrained matter density varies in its gaseous form, and its pressure will vary if constrained.

These two energies are equal at an element's gas transition point.

Lone protons (H$^+$), which constitute 99.97% of all natural hydrogen, can never combine as viscous matter because they have the same electrical polarity. Deuterium (D) and tritium (T), and - theoretically - the proton electron pair (H), can exist in viscous form.

Because orbital radius decreases with increasing temperature, atomic density (and strength) increases with temperature. At low temperatures, atomic density is considerably less than the density of matter.
For example, the density of iron is 7870 kg/m^3
whilst the density of its atom @:
273.15 K: ρ = 0.083888668 kg/m^3
12,412 K: ρ = 7870 kg/m^3

At relatively low temperatures, this *'electron-clouding'* allows lone protons (and positive ions) to share *spare* electron charge capacity in neighbouring atoms, but diminishes with increasing temperature, as electron shells reduce.
I.e. chemical bonding weakens with increasing temperature, which supports the positive-charge premise for molecular atoms (refer to Chapter 4.8)

Stress in viscous matter is actually the pressure required to overcome the magnetic field energy holding adjacent atoms together. I.e. Stress and tensile/compressive moduli are actually pressure, just as their units imply.

Elastic-stress is the pressure necessary to increase the radial separation between adjacent atoms without disrupting the atomic lattice structure.

Plastic-stress is the pressure necessary to separate adjacent atoms whilst disrupting the atomic lattice structure. I.e. adjacent atoms slip from one lattice to the next.

6 Energy

What is energy and how does it apply to the universe we know today?

First; we need to understand that *everything* in the universe comprises packets of electrical and magnetic charge; the interaction of which generates electrical and magnetic fields. It is these and these alone that create all forms of energy in our universe. There is no such thing as gravity, mass, heat, etc.

Energy cannot be lost or gained and it can only be transferred (by electromagnetic radiation) or transformed (from one form of energy to another).

Quanta are packets of electro-magnetic charge.

Energy was not a concept known to Isaac Newton, so he used force to describe energy transfer.
Force is the manifestation of 'energy transferred between two or more bodies separated by physical distance.
This relationship is better known as; Energy = Force x Distance.
The difference between the two concepts can be described thus:
Energy is a force applied over a distance
Force is energy per unit distance

Apart from the non-polar magnetic charge present in all atomic particles, magnetism and electricity are polar; negative or positive. Opposite poles attract and similar poles repel. *Non-polar magnetism only attracts*.

Both types of energy (electrical and magnetic) have static and dynamic counterparts. Because Quanta have opposite electrical and magnetic polarity, when encountering other Quanta, their electro-magnetic energies will *always* be opposite *or* identical according to nature's requirements; i.e. polarity conflict is impossible.

How does this relate to what we see and feel?

Low-temperature scenario: when you see an object, such as a cup, you are seeing all the adjacent atoms in that cup held together with magnetic field energy. In this form, the atoms are sufficiently close together to prevent the atoms in, say, your hand, from passing between the atoms in the cup, allowing you to touch but not penetrate the cup.

The Universe

The weight you feel when you lift the cup, is created by the magnetic energy between the Quanta in the cup and those in the earth.

High-temperature scenario: If sufficient electro-magnetic energy (*heat*) is trapped by the electrons in the cup and your hand, the electric charge energy in all the atoms will exceed the magnetic field energy forcing the atoms in the cup and/or your hand to repel each other and intermingle, in a form that we understand as gas (Dalton's theory).

Because orbiting electrons cannot hold onto energy above a minimum level (e.g. microstates: $N_t = 1.0$; $N_v = 1.5$; $N_p = 2.5$), if an electron traps no further electro-magnetic energy, it will leave its proton partner and electro-magnetic radiation will cease from that proton-electron pair.

Chemistry, and therefore life, occurs when Quanta within a few atoms become detached or are gained through interaction. Thus, the creation of ions and isotopes causes electrical and magnetic disparity between adjacent atoms. Ions react electrically with other ions (chemistry) and isotopes create this capability.

And, finally, how do we perceive these energies:

Mass is the magnetic charge in atomic particles

Light is a particular range of electro-magnetic energy

Gravity is the potential energy between Quanta due to their magnetic charge.

Heat is the electro-magnetic energy emitted by a proton-electron pair

Temperature is the electro-magnetic energy emitted by the proton-electron pair(s) in an atom's innermost shell

All the energy in the universe is generated by friction within satellites by the competing potential and kinetic energies in their force-centres and sub-satellites, stored in neutrons and released during subsequent '*Big-Bangs*'.

All celestial bodies that are without either a force-centre or satellite (for example a galactic force-centre or moon) are black-bodies that generate no internal [heat] energy.

The Universe

The entire universe comprises a fixed, unchanging quantity of energy, it always has done and always will. It was originally contained within the ultimate-body's neutrons, 3% of which was released during the last '*Big-Bang*' and remains unchanged today.
[first law of thermodynamics]

Electro-magnetic radiation is trapped by an electron, converted into kinetic energy, immediately lost to, and radiated by, its proton. If insufficient electro-magnetic energy is trapped by an electron, it naturally leaves its proton-electron partnership and continues in free-flight at the linear and angular velocities it had when it left its proton. However, electrons never stop moving.
[second law of thermodynamics]

The natural (minimum entropy) state of the universe is the reversion to protons and electrons
[third law of thermodynamics].

The Universe

6.1   Potential

Potential energy is the direct (straight-line) attraction or repulsion between quanta due to their electrical or magnetic charges, or their electrical and magnetic fields.

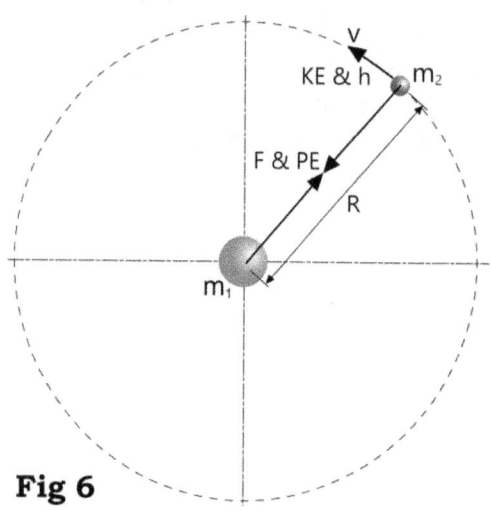

Fig 6

The attraction between all matter - what we currently understand as gravity - is actually due to the non-polar magnetic charge in all quanta.

Electrical charge, and electrical & magnetic fields are polar. In all cases, like poles repel and unlike poles attract.

Whilst potential energy due to electrical or magnetic charge permeates throughout the universe, its influence due to electrical and magnetic fields, whilst stronger, is limited to within their field lines.

In a balanced system (e.g. orbits): gravity (magnetic charge) and centrifugal are potential energies; negative and positive respectively, both of which are always equal. They are calculated like this:

$$PE = m.g.R = G.m_1.m_2/R$$

In the special case of circular orbits, because the potential energy between a force-centre and its satellite is always twice the satellite's kinetic energy, it may be calculated like this:

$$PE = 2 \cdot \tfrac{1}{2}.m.v^2 = m.v^2 \text{ (Henri Poincare)}$$
where 'v' is the satellite's velocity and 'm' is its mass.

6.2 Kinetic

Kinetic (dynamic) energy, which exists in all moving particles, is always positive and induced via electrical, magnetic, electro-magnetic or impact [potential] energy.

Kinetic energy only applies to bodies of mass; therefore, it cannot apply to electro-magnetic energy (EME), because EME has no mass.

In the case of an electron that is partnered with a proton (a proton-electron pair) EME is absorbed by the orbiting electron which it converts to kinetic energy.

The kinetic energy of a satellite in a non-circular elliptical orbit is induced by the varying potential energy between the satellite and its force-centre. Whereas, the kinetic energy of a satellite in a circular orbit must be supplied by the satellite itself. In the case of a man-made satellite, this is usually provided by solar radiation; EME. In the case of an orbiting electron, this is supplied by the EME in the electron's environment. It is calculated like this:

$$KE = \tfrac{1}{2}.m.v^2$$

6.3 Electrical & Magnetic

Electricity and magnetism are essential converses; one cannot exist without the other.

Whilst attraction between magnetically charged particles is weaker than that between electrical charges, electrical energy is shared and magnetic energy accrues. Therefore, as the number of particles increase, magnetic interaction will increase and electrical interaction will decrease. This is the reason magnetism (gravity) is so strong between celestial bodies and electrical attraction is negligible.

The difference between magnetic and electrical particle charges is **the coupling ratio**: φ = 4.40742E-40

Relative rotary interaction between electro-magnetically charge particles generates polar electro-magnetic field energy.

Electrical field forces travel from negative towards positive and magnetic field forces travel in the opposite direction.

The electrical and magnetic charges in every particle are felt by every other particle in the universe, according to the general formula;
$$F = K.v_1.v_2/A \; \{N\}$$
Where; F is the electrical or magnetic force between two charges; K is a constant; v_1 and v_2 are variable charges and A is the spherical surface area at the radial distance between them.

Because electrical energy is shared, electrical force should look like this;
$$F = K.(v/R)^2 \; \{N\}$$
where v is the lesser of the two variables.

You will have seen the above formula written as follows;
$$F = K.v_1.v_2/R^2 \; \{N\}$$
giving the impression that force and energy diminish with the square of the distance between interacting bodies; contradicting the conservation of energy and the first law of thermodynamics, which must be incorrect!

In reality, force & energy do not vary with radial distance, they are simply distributed over the spherical area (A) at radius 'R'.

The Universe

6.4 Electro-Magnetism

Electro-magnetic energy is generated (and radiated) by proton-electron pairs and is the means by which energy is transferred. It always travels at the same velocity (299792459 m/s), which we refer to as the *speed of light*, but is of course, the same speed for *all* electro-magnetic energy.

Electro-magnetic energy is susceptible to magnetic charge energy in a planet or star and will be deflected by it according to Newton's gravitational constant (G)

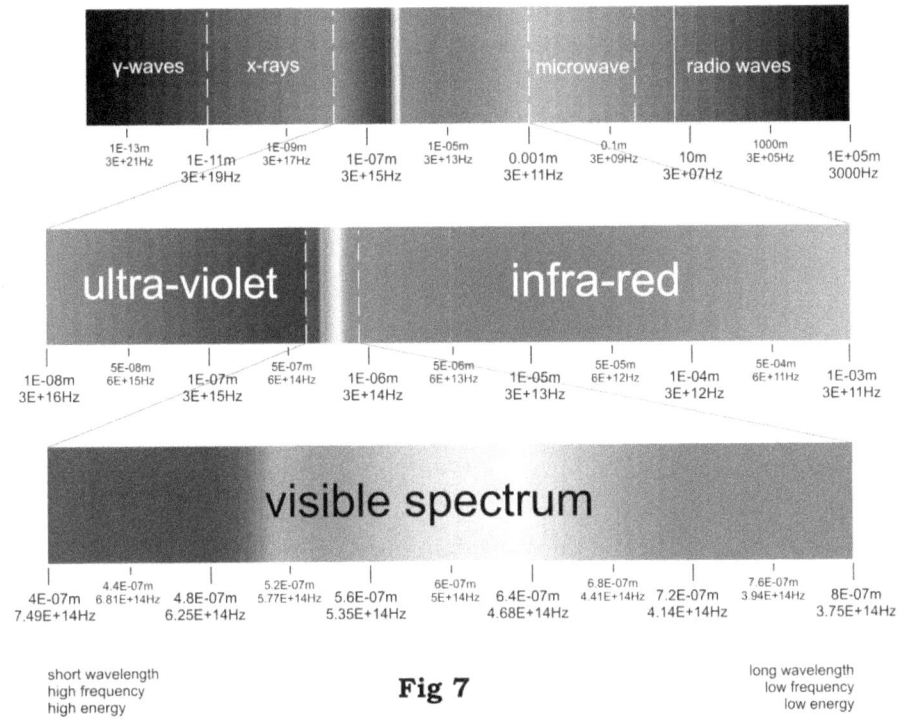

Fig 7

The bands of electro-magnetic energy (light, radio, X, γ, etc.) are defined by the orbiting electron's kinetic energy (velocity) and radiated in different wavelengths that we perceive as light, heat, etc. The magnitude of a radiated energy wave: its brightness, temperature, etc. is defined by its frequency. It possesses both electrical and magnetic polarised energy but no *mass*.

The proton electron pairs in all elements generate and radiate EME directionally according to their lattice structures.

Calcium, for example, is an atom with twenty proton-electron pairs and will therefore emit electro-magnetic energy in twenty directions. Millions of such atoms will emit electro-magnetic energy in millions of directions.

Because the electro-magnetic energy radiated by the innermost proton-electron pairs within a body gets bounced around by its neighbouring pairs, it will take much longer to escape to atmosphere. This is why the surface of a body cools faster than its centre.

Electro-magnetic radiation is deflected as it passes a large body e.g. a planet or star because of non-polar magnetic charge, i.e. magnetism.

Heat is what we feel from the electro-magnetic energy emitted by proton-electron pairs. Temperature is how we measure this heat. Each atomic shell will emit heat at a different temperature, the highest being that emitted from the innermost shell and that which we measure.

6.4.1 Wave Characteristics

Electro-magnetic radiation travels in waves with a frequency, wavelength and amplitude commensurate with its energy (temperature). Its energy is what we understand (feel) as heat. Heat is the collection of energies radiated by proton-electron pairs, each of which will be at a different level dependent upon the orbital velocity of the electron responsible for it. The highest energy is emitted by a pair with its electron in the innermost shell and the lowest by a pair with its electron in the outermost shell.

In many publications, you will see the electro-magnetic wave depicted as described in Fig 8;**A**. The problem with this configuration is that the kinetic energy (E_T) of the electron must vary between 0 and 2.KE in each cycle for it to work, which is impossible. The kinetic energy in an orbiting electron remains constant throughout each cycle. The shape of the wave, therefore, must be as described in Fig 8;**B**, in which the total energy (E_T) in the wave is constant. I.e. when the electrical component is at a maximum amplitude, the magnetic amplitude is zero, and vice versa.

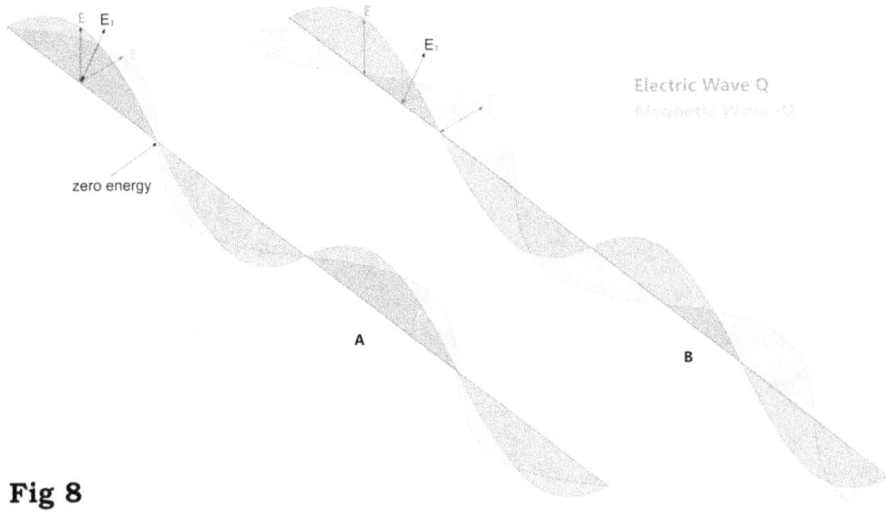

Fig 8

Nobody knows what an electro-magnetic wave actually looks like, so it could be similar to that shown in Fig 9, i.e. a helical wave that varies between maximum/minimum positive to maximum/minimum negative electrical and magnetic energy as it helically rotates. In fact, given the energy source (an orbiting electron), this option is more likely than that shown in Fig 8;**B**

Fig 9 shows a helical electro-magnetic wave winding its way through;

$+E_e$; $-E_m$; $-E_e$; $+E_m$. Measuring either electrical or magnetic energy alone will produce a flat sinewave profile such as that shown, but it always emits the total kinetic energy in the electron (E_T).

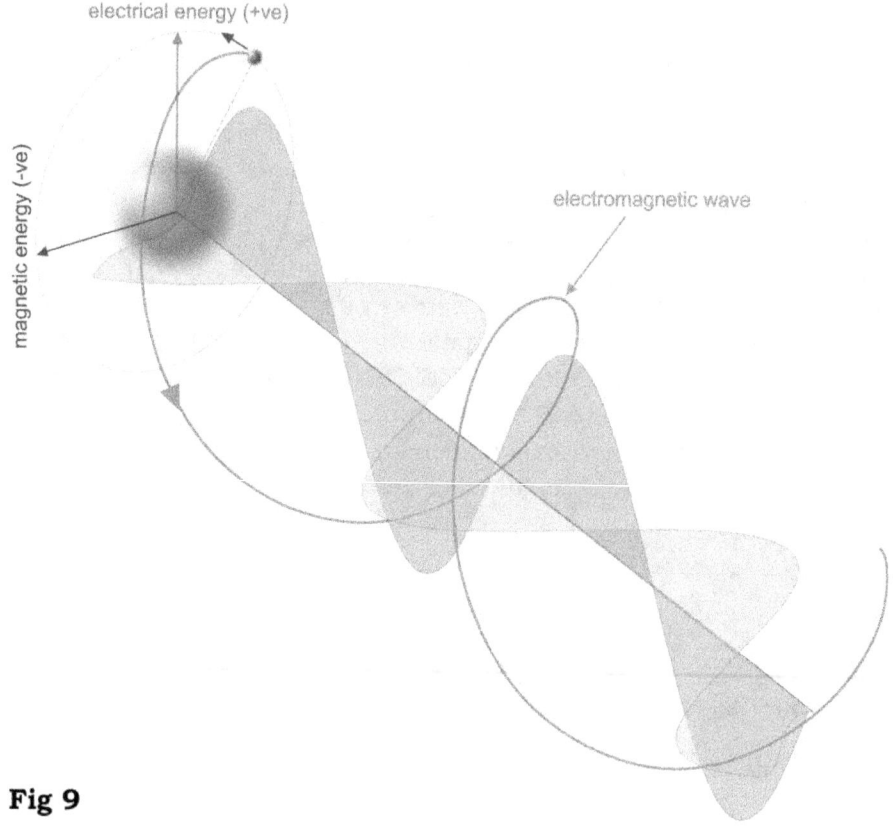

Fig 9

You will see many claims that amplitude increases with energy. This is incorrect. Whilst his *formula* (h) was incorrect, Max Plank realised that energy is proportional to frequency, not amplitude.

The electro-magnetic energy produced by each proton-electron pair will be different to that produced by a proton-electron pair in the same atom but in a different shell. The difference will be consistent with the kinetic energy in the electron generating the energy and is what gives the different atomic numbers their characteristic Balmer lines.

Irrespective of wave shape, electro-magnetic energy possesses the [kinetic] energy transferred by the orbiting electron and is emitted by the electrical charge in the proton (e'), which varies between electrical and magnetic energy according to equal sinewave profiles normal to each other.

6.4.2 Heat & Light

It is important to remember that all the electro-magnetic energy generated in the universe is just that; electro-magnetic energy (EME). It possesses; no mass, light, heat or sound – nothing, apart from energy.

If you or I, devoid of electrons – impossible I know, but bear with me – were to sit in the space between the sun and the earth, we would not be able to detect the sun's radiated EME. It would be invisible in every sense to the fictitious you (or me). EME is useless to all forms of life unless it can be detected.

Whilst EME doesn't deteriorate with distance travelled, we don't feel the sun's surface temperature (5788K) here on Earth because the energy *density* (Joules per square metre) radiated at the sun's surface is distributed over a spherical surface area between 45,000 and 48,000 times greater (dependent on the time of year). Therefore, the EME *density* we receive will be correspondingly less.

Life here on Earth, has evolved to detect and use this energy through our complex molecules. The trouble is, such molecules have energy tolerance levels, outside which they would no longer function; i.e. their state-of-matter strength or condition (gas-viscous) could change, or inter-atomic bonding could fail.

For example; if a block of viscous iron received sufficient EME to increase its proton electrical charge energy above that of its atomic magnetic field energy, it would become a gas. And it would cease to be *a block of iron*.

There are tolerance levels regarding acceptable amounts of EME any living organism can receive and remain functional. Therefore, all living organisms have developed senses, that can be used to ensure that these tolerance levels are maintained.

We (humans) have five senses - if you exclude time - smell, touch, taste, sight and hearing; each of which were developed for application and protection.

6.4.2.1 Heat

Temperature is a measure of the kinetic energy of an element's innermost orbiting electrons.

Heat is the kinetic energy in an atom's electrons. The heat we *feel* is from the senses we have developed to tell us when this kinetic energy is too high or too low. Electron kinetic energy is generated by the EME it absorbs from its surroundings.

You can't damage a block of iron, so it doesn't need senses. It doesn't matter how many times you change it from gas to viscous and back again its atoms will always remain iron. The higher its temperature the stronger its atoms remain, until the innermost electrons achieve the *speed of light*, when they will become a different element (Z-1 or Z-2).

All the EME in our environment is shared between all of our electrons. The greater the EME *density*, the greater the *heat* we feel. Irrespective of the *temperature* of the atoms that generated the EME, if the energy *density* can be shared throughout all the electrons in our body without exceeding its tolerance levels, we will remain functional.

In other words; it is not the *temperature* of the atoms emitting the EME that defines our body-temperature; it is defined by the quantity of EME absorbed by our body's electrons.

The highest possible temperature in nature is 623316124.717178 K, which is the temperature of a proton-electron pair immediately before it becomes a neutron. This is therefore the temperature at the core of bright stars, where all neutrons are created.

The lowest possible temperature that can be generated by a proton-electron pair occurs immediately before the electron leaves its proton partner. This is referred to as the 'cold' temperature; 2.04274907568265 K.

All Heat Is Radiated:

Conduction *is the transfer of EME between the electrons in adjacent atoms in viscous matter.*

Convection *is the movement of atoms (and molecules) to a position where the electrical charge repulsion energy between atomic protons can balance in three-dimensions, and where it can transfer this energy to atoms with less heat. It only occurs in gases that are under the influence of potential energy.*

6.4.2.2 Light

Question: "If colour is defined by EME, and the sun's surface temperature is 5778K and looks yellow; how can my towel (at 300K) also look yellow?"
Not an easy question to answer, but I'll try ...

Colour is a range of EME wavelengths (4E-07m > 8E-07m) that we cannot detect until it is absorbed by the electrons in our optical receptors (eyes). **Light** is simply EME intensity, or put simply; the number of electro-magnetic rays per square metre.

We (humans) have developed eyes to detect a bandwidth of EME that best suits our purpose. Other lifeforms have developed receptors that best suit their own environment which may be outside the aforementioned *optical* range. Irrespective of a lifeform's preferred optical bandwidth, the purpose of sight is the same; to see what's in its environment and how best to exploit it.

Every proton-electron pair in the universe, *including that block of wood in the garden*, emits EME at a wavelength commensurate with the kinetic energy in its electrons. But you cannot see the EME radiated by that block of wood here on Earth, because it is radiated in the infra-red range. This is the reason infra-red cameras reveal objects in the dark here on Earth. They are actually capturing the electro-magnetic energy given off by the objects themselves, not the electro-magnetic energy radiated by the sun.

Unlike in a prism, the diffraction of light through natural matter is not organised. The image of that block of wood, is the sun's EME reflected and/or refracted by or through its constituent atoms and molecules.

Just as with heat, if the EME received by your eyes is not too intense, i.e. its density remains within your body's tolerance levels, the sun's rays you see will do you no harm.

Whilst our sun's *surface* comprises proton-electron pairs as hydrogen and helium due to its heat, its internals comprise all of nature's elements. It therefore emits a preponderance of electro-magnetic radiation at all the wavelengths from yellow to gamma. All the infra-red, micro-wave and radio-wave energy coming to us is generated by colder celestial bodies in the universe (force-centres with few or no satellites).

So, when you see a *yellow* towel here on earth, your eyes are detecting the EME radiated by the sun at a wavelength of ≈6.3E-07m diffracted by the molecules in your towel, but at an intensity that will not harm your eyes.

6.5 $E=mc^2$

$E=mc^2$ was first postulated by Henri Poincaré towards the end of the 19th century ($c = \sqrt{[E/m]}$), however, he did not explain its physical relevance other than he believed it represented a terminal velocity.

During the creation of this publication, I discovered that $E=m.v^2$ represents the *potential* energy between a satellite and its force-centre in circular orbits (e.g. atoms), where; PE = -2.KE
KE = ½.m.v²; PE = m.v²
and at the speed of light, an electron orbits its proton at radius R_n, which is a fundamental constant where PE = m.c² at the creation of a neutron.

Whilst Poincaré's formula ($E=m.c^2$) does indeed refer to a terminal condition, it refers to the ultimate potential energy between a proton and its electron partner that is *orbiting* at 'c'. And occurs when the magnetic attractive field energy exceeds centrifugal repulsion energy and the electron combines with its proton to create a neutron. It does not refer to an electron in free-flight travelling at the speed of light, or in fact, kinetic energy of any kind.

That said, electro-magnetic energy can only be radiated whilst an orbiting electron is travelling at less than 'c', which means that no electron can *naturally* achieve this kinetic energy; E = ½m.c², which it gets from electro-magnetic radiation; E = h'/A (where amplitude; A = R_n). However, this limitation *does not* mean that an electron, or anything else, cannot travel faster than 'c' if given sufficient energy artificially.

This means that in Relativity, $E=mc^2$ has been inappropriately applied to kinetic energy to describe mass-energy variation with velocity, which does not actually occur; mass does not vary with velocity and there is no such thing as mass.

Combining the theories from Newton, Planck and Poincaré:
Assuming 'm' is a unit of mass of ultimate density: m = ρ_u
    Newton: G = $a_o.c^2$ / m (Keith Dixon-Roche)
    Planck: F = c^4/G
    F = $m.c^4$ / $a_o.c^2$
    F = E/R
    E = F.a_o
    E = m.c² (Poincaré)

All of which demonstrates that 'E=mc²' applies to potential, not kinetic energy. Therefore, electrons (and all matter) in free-flight are not limited to the speed of electro-magnetic energy (e.g. light).

E=mc² has nothing to do with kinetic energy and mass does not change into energy with increasing speed.
E=mc² (which was discovered/prophesied by Henri Poincaré) refers to circular orbits where potential energy (PE) is twice kinetic energy
(PE = 2 . ½.m.v²).
At the velocity of light (v = c), a proton and its electron combine to create a neutron.

'E=mc²' refers to potential energy, not kinetic energy.

The Universe

7 The Universal Machine

The universe is essentially an energy generator and storage facility.

Its mechanical components include all of its celestial bodies together with all the Quanta they contain. Every single particle in it is necessary to make it work.

Its operational processes include;

    1) Orbits

    2) Spin (friction)

    3) Core Pressure

7.1   Orbits

An orbit is the path followed by a satellite about its force-centre. This path is *always* elliptical, just as Kepler told us. It looks like the image in Fig 10

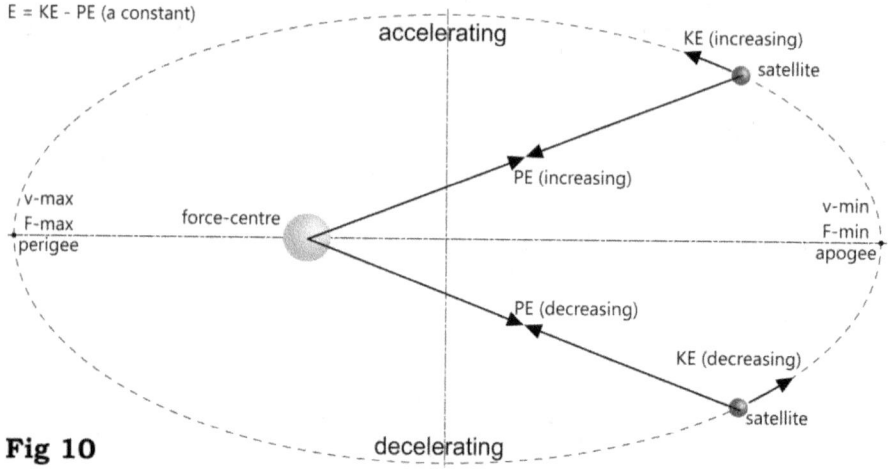

Fig 10

Potential energy traps a comet as it passes by a force-centre, that then becomes a satellite.

As a comet passes a force-centre, the point at which it becomes a satellite is determined by the relationship between their potential energy and the comet's kinetic energy, which must comply with the force-centre's constant of proportionality. If these conditions are not met by a comet, it will not become a satellite.

Orbital shape and period is defined only by a force-centre's mass, according to the formulas: $m = (2\pi)^2 / G.K$ & $K = t^2/a^3$
Where:
'm' is the mass of the force-centre
'G' is Isaac Newton's gravitational constant
't' is the orbital period
'a' is half the length of the elliptical major axis
'K' is the orbital constant of proportionality'

It is important to remember that whilst the kinetic and potential energies vary throughout an orbit, the *total* energy (E) in the orbital system never changes. Potential energy is negative and kinetic energy (velocity) is positive; **'E = PE + KE'** is a constant.

The constant of proportionality is identical for every satellite orbiting the same force-centre.

An orbit only works for a perfect ellipse, which is how we know that space and time do *not* deform around celestial bodies (Relativity).

The perpetuality of a satellite orbit only works in a vacuum, which is how we know that there is no dark matter.

Circular orbits are special in that their:

satellites must provide their own kinetic energy

and

system potential energy is always exactly twice the magnitude of the orbital kinetic energy

This orbital system was discovered and explained by Kepler, Galileo and Newton over 300 years ago; every single aspect of which is still perfectly valid today. Despite what you may hear to the contrary; it even works for satellites orbiting at well above the *speed of light*.

The Universe

7.1.2 Terminology

An **orbit** is the path followed by a satellite around its force-centre. For example, our moon is in orbit around our Earth (a planet), which is in orbit around our sun (a star), which is in orbit around Hades (a galactic force-centre). Electrons orbit their protons (a proton-electron pair).

An orbiting body (or mass) is referred to as a **satellite** and the body about which it orbits is referred to as its **force-centre**.

These orbital systems have group names such as:

Solar Orbit: A star's orbital path around its galactic force-centre
Planetary Orbit: A planet's orbital path around its star
Lunar Orbit: A moon's orbital path around its planet
Atomic Orbit: An electron's orbital path around its proton

Collectively, everything (including the force-centre) orbiting a ...
... galactic force-centre is called a **galaxy**
... star is called a **solar system**
... planet is called a **lunar system**
... proton is called a **proton-electron pair**

An orbit is *always* a perfect ellipse, exactly as Johannes Kepler stated. An ellipse can be any two-dimensional (flat) elliptical shape including a circle.

There is a major difference between a genuine elliptical orbit, i.e. one in which its axes *are not* identical in length, and a circular orbit, i.e. one in which both its axes *are* identical in length:

Satellites following non-circular **elliptical orbits** (e.g. stars, planets, moons, comets, etc.) keep going because of the potential and kinetic energies (Newton's constant of motion) between a satellite and its force-centre.

Satellites following **circular orbits** (e.g. electrons, communication, spy, etc.), keep going because they provide their own kinetic energy.

Mass is magnetic charge

Gravity is the potential energy radiated by magnetic charge

Velocity in orbits refers only to the curvilinear motion of a satellite in its path around its force-centre. It does not refer to rotational (angular) motion in a satellite or its force-centre

The Universe

Force is energy per unit distance

Energy is a force applied over a specified distance

Kinetic energy is the energy in a satellite due to its velocity

Potential energy is the attractive/repulsive energy between a satellite and its force-centre

Planetary Spin; the angular velocity (radians per second) in a body rotating about an axis that passes through its centre of mass

7.1.3 Orbital Laws

The laws of orbital motion are the mathematical formulas that describe the properties of a satellite's curvilinear motion around its force-centre. It is important to understand that rotary motion (spin) plays no part in Isaac Newton's laws of orbital motion. However, the laws describing spin in a satellite and its force-centre may be derived from them.

The Principal Orbital Laws are as follows:

Every Orbital system must comprise only one force-centre and at least one satellite

The shape and period of every orbit is defined by the mass of its force-centre (constant of proportionality)

Changing the properties of a satellite will not alter its orbital shape or period

The potential energy in circular orbits is always twice the satellite's kinetic energy

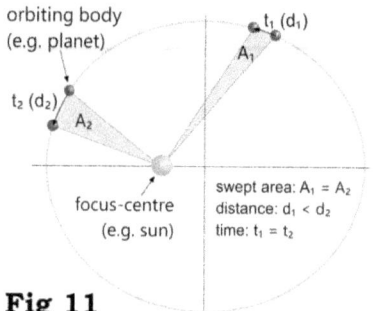

A useful tip from Kepler that was later verified by Newton is that the relationship between the swept area inside the ellipse for any given period of time of a satellite's orbit will always be identical (Fig 11).

Fig 11

Isaac Newton's mathematical laws were complicated and incomplete. Like all scientists, Newton apparently didn't like to give away all his secrets. He actually omitted a few crucial features making it difficult to fully understand his work.

Having thoroughly evaluated his orbital theory, I have managed to simplify his mathematics and fill in all the gaps. It is now, therefore, possible to calculate everything you need to know about an orbital system using just a few simple formulas.

The Universe

Whilst this book is not intentionally mathematical, I have considered it prudent to list these formulas for your information.

Sym	Description	units
t	Orbital period	s
R^P	Radius at the orbital perigee	m
θ	Any angle in orbit from apogee	c
R^A	Radius at the orbital apogee	m
m_2	Satellite mass	kg

Table 7-1: *Input Data*

Sym	Formula	Description	units
R	$p / [1 - e.\cos(\theta)]$	Orbital radius at θ	m
a	$(R^P + R^A) / 2$	half the major axis of the ellipse	m
e	$-R^P + \sqrt{[R^{P2} - 4.a.(R^P-a)]} / 2.a$	eccentricity of the ellipse	
b	$\sqrt{[a^2.(1-e^2)]}$	half the minor axis of the ellipse	m
p	$a.(1-e^2)$	half-parameter (of orbital path)	m
f	R^P	focus distance (orbital perigee)	m
x'	$a - f$	distance from focus to ellipse centre	m
A	$\pi.a.b$	orbital swept area	m²
L	$\pi.\sqrt{[2.(a^2+b^2) - (a-b)^2/2.2]}$	orbital path length	m
K	t^2/A^3	orbital constant of proportionality	s²/m³

Table 7-2: Orbital Shape

Sym	Formula	Description	units
m_1	$\varphi.(2\pi)^2 / G.K$	Force-centre mass	kg
m_2	input	Satellite mass	kg

Table 7-3: Masses

Sym	Formula	Description	units
v^P	$2.A / t.R^P$	satellite velocity at orbital perigee	m/s
v, v_c	$2.A / t.R$	satellite velocity at θ	m/s
v^A	$2.A / t.R^A$	satellite velocity at orbital apogee	m/s
g^P	$-v^P.v^A / R.(1+e)$	gravitational acceleration at perigee	m/s²
g	$-v.v^A / R.(1+e)$	gravitational acceleration at θ	m/s²
g^A	$-v^P.v^A / R.(1+e)$	gravitational acceleration at apogee	m/s²
F	$-g.m_2$	gravitational force from force-centre	N
F_c	refer to Chapter 3.2.7	centrifugal force in satellite	N
PE	F/R	potential energy between bodies	J
KE	$½.m_2.v^2$	kinetic energy in satellite	J
E	$PE + KE$	total energy	J
h	$R.v$	constant of motion	m²/s

Table 7-4: Orbital Performance

The above formulas were not only used to calculate the orbital properties of over 200 of our solar system orbits, all of which were verified using NASA data, they were successfully used to analyse our orbit within the Milky Way galaxy and the atom. Newton's system is indeed correct and universal.

7.1.3.1 Goodricke & Algol

In 1784, John Goodricke discovered the [supposed] binary nature of the star originally named Algol. As one of the binary stars passed in front of the other, their combined brightness dimmed, revealing two important facts:

1) One of the stars was bright (hot) and the other was dark (cold)

2) Only the bright star was a force-centre for an orbital system

The heat generated by the *bright* star is due to its dedicated satellite population. The other darker star (or large planet) is actually in orbit around the bright-star but has few satellites of its own. In which case, the dark partner generates no fissionable energy.

Once a star (force-centre) has acquired its satellites, it can trap others, or even a twin but it will not share its satellites.

This discovery also reinforces the fact that stars generate their own internal heat from their satellite population and the proton-electron pair behaviour in atoms.

7.1.4 Station-Keeping

Satellites are often but temporarily influenced by other celestial bodies. For example, Fig 12 shows a planet orbiting its sun but close to another body outside its orbit.

Kepler believed that the potential energy between the satellite and the external body will pull them together, out of their respective orbits temporarily altering their orbital paths. He believed that the satellite will accelerate as it travels towards the external body and decelerate (relative to the external influence) after passing it. However, these variations are effectively cancelled out as a result of Kepler's and Newton's 'equal time swept-area' law and the conservation of energy.

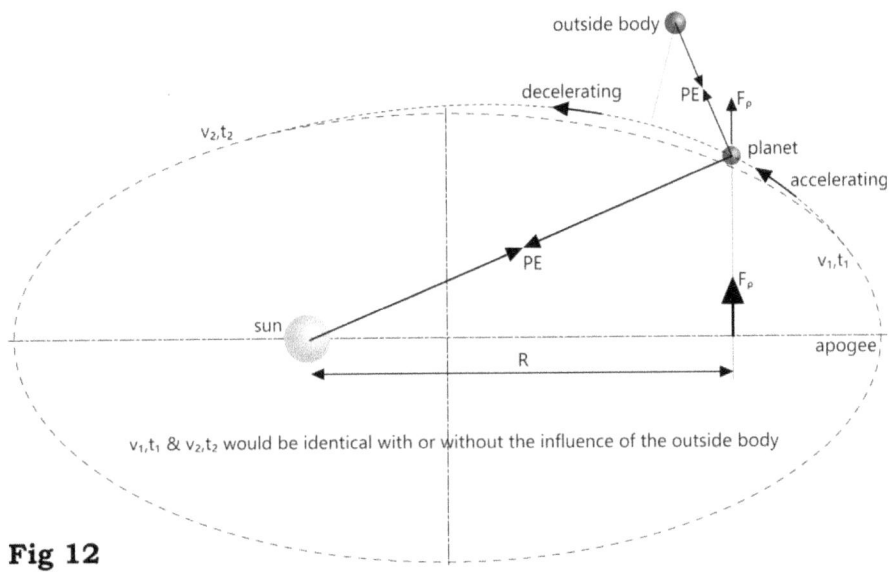

Fig 12

This is, however, incorrect. In order for a satellite to remain in orbit, its centrifugal and centripetal forces must be equal, so it cannot leave its orbit. What actually happens is that a torque ($T_p = F_p \times R$) is applied to the satellite's orbital axes causing a gradual rotation (of the orbit) about its force-centre. The frequency and magnitude of such influences define the rate of orbital precession.

This orbital rotation is responsible for occasional impacts between adjacent satellites, creating galactic and solar comets; and asteroid belts.

7.1.4.1 Centrifugal Force

Any orbiting mass will be subject to a centrifugal acceleration (a) that must always balance with the acceleration (g) induced by gravity.

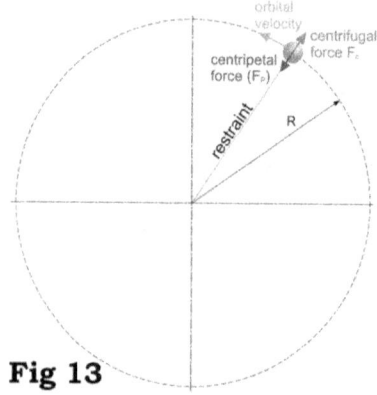

Fig 13

If you swing a ball - tied to a length of string - around your head, *centrifugal* force is pulling the ball away from you. But it also induces a tensile force in the string, pulling the ball towards you. This is *centripetal* force, but it is also *potential energy*; the equivalent of gravity.

Christiaan Huygens gave us the mathematical relationship between this and its velocity in a circular orbit (Fig 13); $a = v^2 / R$, where 'v' is its curvilinear velocity and 'R' is its orbital radius.

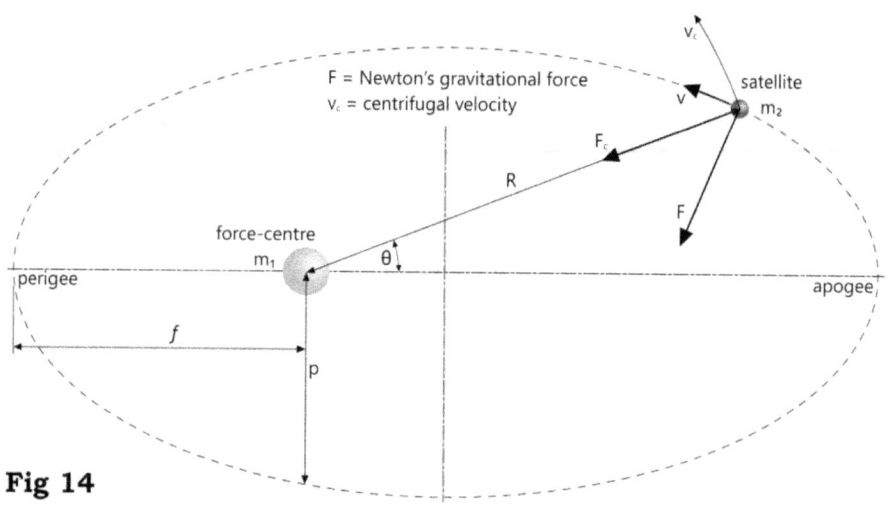

Fig 14

However, the above velocity (v) must be modified for elliptical orbits (v_c) as shown in Fig 14. Its magnitude is dependent upon the orbital eccentricity and varies with orbital radius (R).

7.1.4.2 Return to Station

When a displacement force tries to pull a satellite off course, a restoring force will maintain it *exactly* in its orbital path, where both centrifugal and gravitational acceleration balance; i.e. where they are equal and opposite.

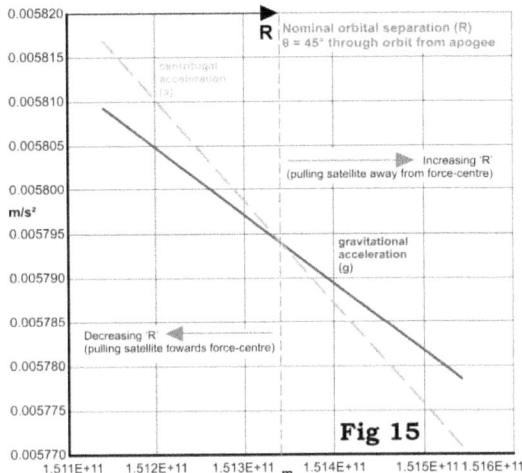

Fig 15

Fig 15 provides a graphical representation of the restoration force on the earth 45° through its orbital path from its apogee.

As the earth is pulled away from the sun (increasing R), gravitational acceleration (g) increases faster than centrifugal acceleration (a), pulling the earth back towards the orbital path when the displacement force is released.

As Earth is pulled towards the sun (decreasing R), centrifugal acceleration (a) increases faster than gravitational acceleration (g), pulling the earth back towards the orbital path when the displacement force is released.

As you can see (Fig 15), *exact* balance occurs at the orbital path: at nominal orbital separation 'R'.

This relationship between centrifugal and gravitational acceleration is what maintains the orbital path in the event a satellite comes under the influence of external forces.

The above process requires a perfect elliptical orbit, *exactly* as Kepler defined.

7.1.4.3 Relativity

Relativity states that space-time deformation around force-centres alter satellite velocity, e.g.; $v = v/\sqrt{[1+(c/v)^2]}$.

Not only is this poppycock ($c \neq c/\sqrt{2}$), it also causes station-keeping to fail; i.e. planets pulled out of their orbit would not return (Fig 16). Because we know that they do, Relativity must be wrong.

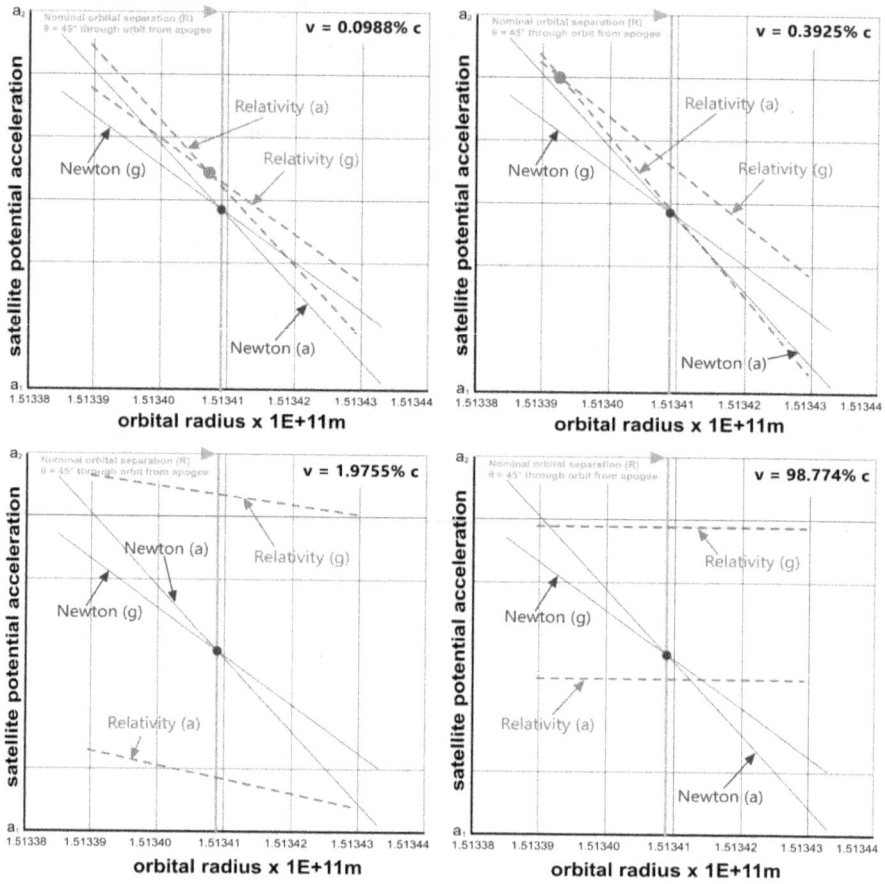

Fig 16

Elliptical orbits are an indisputable fact of nature. This has been repeatedly demonstrated since Kepler's discovery in the 17th century. Its mathematical laws show that an exact ellipse is *fundamental* to the constant of [orbital] motion and thereby essential to maintain satellite paths in non-circular orbits. Relativity requires a distortion of this ellipse, rendering the orbital laws unworkable; yet we know that Newton's universal orbits work perfectly irrespective of size, shape and velocity.

It must therefore be concluded that Relativity is not only unnecessary; it must also be incorrect.

7.1.5 Orbital Planes

The spin induced in a force-centre by its orbiting satellites, influences the orbital plane of its satellites (Fig 17).

Fig 17

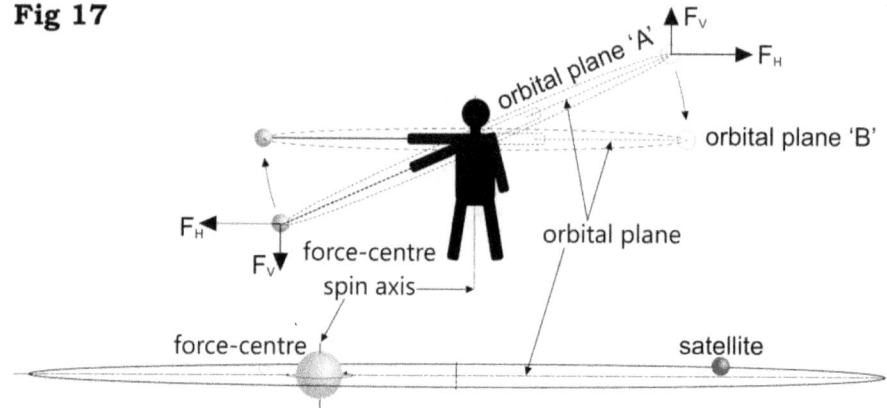

The rotational kinetic energy in a force-centre naturally causes the orbital planes of its satellites to settle at 90° to its spin axis. This phenomenon can be demonstrated by attempting to swing a ball about *orbital plane 'A'* in which 'F_v' is non-zero. The orbital plane will always settle at the lowest energy condition; *orbital plane 'B'*, where 'F_v' is zero. The same forces are at work in planetary orbital systems.

7.1.6 The Importance of Orbits

Everything in the universe, without exception, is dependent upon orbits.

Without them, there would be no heat, elements, chemistry, life, electro-magnetic energy or matter. In fact, the entire universe would be a cold, dark sea of Quanta.

Orbits are the source of *all* universal energy; through spin (friction) and neutrons.

So, it is not surprising that Newton's laws of orbital motion are without doubt *the* most important mathematical laws in science. They explain how *everything* in the universe works, including the atom!

Without the electron orbit, there would be no atoms. There would be no elements and therefore no molecules. Neutrons would not exist. Electrons would vanish into infinity, leaving behind a sea of lone protons.

If you were to remove the lunar orbit, e.g. the earth's; there would be no differential core-mantle rotation and therefore, no planetary heat or magnetic field. Days would be excessively long (one day would be longer than a year). There would be no continental drift, no atmosphere, no tides, no weather, no volcanic activity, etc., and therefore no life as we know it.

If we were to lose the planetary orbit, we would have no seasons, no rotating sun; one face of the planet would remain permanently hot whilst the other would remain permanently cold.

If we lost the galactic orbit, our sun would not be generating its internal heat or light; i.e. there would be no stars and hence no neutrons and therefore no subsequent 'Big-Bang'!

7.2 Spin

Spin is the source of all universal energy through internal friction. It applies to all planets and all stars.

All astrophysicists today claim that either there is no force spinning the planets or that it is impossible to calculate owing to the chaotic nature of the solar system. Both these views are incorrect. Planetary spin can be easily predicted with the same degree of accuracy as Newton's laws of orbital motion from the potential and kinetic energies calculated using them. This same theory also applies to galaxies and the Atom.

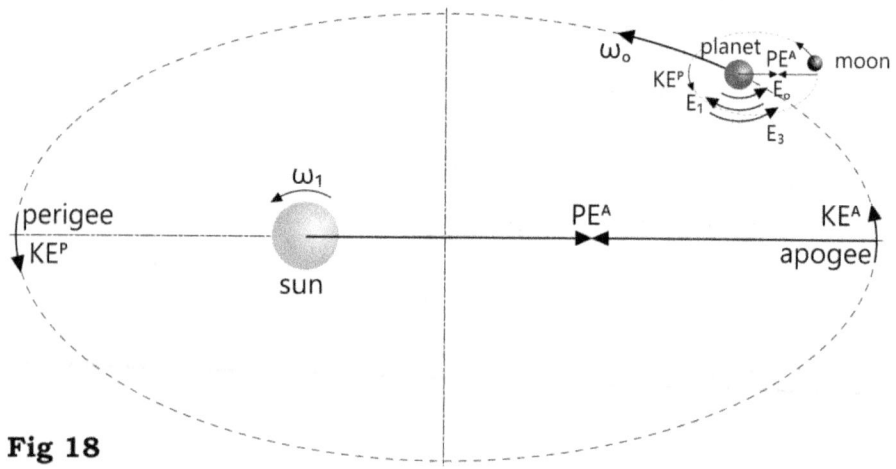

Fig 18

Fig 18: Potential energy between a force-centre and that of its satellite will naturally cause the satellite to spin at the same angular velocity and in the same rotational direction (prograde) as its orbit around the force-centre (ω_0). However, spin energy induced in the satellite by the force-centre's own rotational kinetic energy will cause the satellite to rotate in the opposite (retrograde) direction ($-\omega_1$).

If a planet's moon is orbiting in the same rotational direction as the planet's orbit e.g. prograde, the moon's kinetic and potential energies will cause the planet to rotate also in a prograde direction (ω_1).

Only the potential and kinetic energies in a force centre, its orbiting satellite(s) and their secondary satellites induce spin in each other. It is not as difficult or complex as everybody appears to believe.

For example: Venus's opposite rotational direction cannot, apparently, be explained, but the reason for it is quite simple once you understand spin theory.

E_3, which is generated by a planet's moon(s), is always considerably greater than E_1 and E_0 (Fig 18) and therefore defines a planet's rotational direction. As E_3 in a planet with no moon is zero its orbital energies (E_1 & E_0) determine its spin direction. Being a (relatively) large planet, E_1 is the dominant factor in Venus's spin direction.

The same argument applies to Mercury except in its case its smaller size means that E_0 is dominant so it spins in the same direction as the other planets in our solar system.

A significant difference will occur between the angular velocity of a planet's core and its mantle only if it has a substantial moon.

The only unknowns you need in order to calculate planetary spin are the polar moment of inertia of the planet and that of its sun. When calculating the spin in a planet, you need either its angular velocity (ω) or its radial modifier (Δ). If you have one, you can calculate the other.

The competing rotational energies (E_1, E_3 & E_0) along with their different distribution (@ the core or throughout) are responsible for the conflicting angular velocities in a planet's core and mantle and thereby generating its internal heat. This velocity difference in the earth is also responsible for generating the earth's magnetic field. The positive value of $\delta\omega$ in the earth means that the right-hand rule depicts magnetic north in the correct direction.

Gas planets exist as such because they have been able to attract sufficient satellite *mass* to melt their [surface] crusts through internal friction. Unlike a star, however, they cannot generate the heat required to create neutrons. It is also probable that the heat lost to a gas planet's heavy surface gases will form a surface skin.

This calculation method is as accurate as Newton's own laws of motion and is essentially an extension of them. Therefore, not only is planetary spin predictable, it is both simple and accurate.

The Universe

The component spin energies in a celestial body are summarised as follows:

E_0 is the natural rotational energy developed in a satellite, assuming it presents the same face to its force-centre. It may be represented by the rotation that would be expected in a ball swung about your head, at the end of a length of string. The direction of spin induced in a satellite by 'E_0' will be the same as that of the satellite's orbit. The spin-period will be identical to the satellite's orbital period.

E_1 is the rotational energy induced in a satellite by the rotational energy in its force-centre. This rotation may be represented by a pair of gears and behaves in the same way; i.e. the direction of spin induced in the satellite by 'E_1' will be opposite to that of the satellite's orbit.

E_2 is the final (total) spin in the satellite; i.e. the sum of all the other rotational energies.

E_3 is the spin energy induced in a satellite by its own sub-satellite(s). The direction of spin induced in the satellite by 'E_3' will be the same as the orbital direction and kinetic energies in these sub-satellites.

7.2.1 Polar Moment of Inertia

Polar moment of inertia of an homogeneous mass is calculated thus:
$J = \tfrac{2}{5}.m.r^2$
Where 'm' is the mass of the satellite and 'r' is its outside radius (Fig 19).

However, this formula only works as shown if the body comprises a mass of constant density, which is not the case for celestial bodies, such as; planets, stars, moons, etc. Potential energy tends to ensure that the denser matter migrates towards their cores.

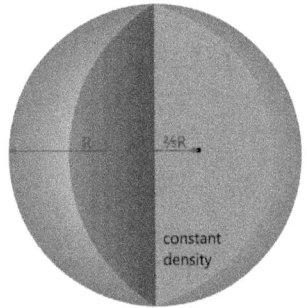

Homogeneous Sphere

This problem can be solved by using a radial modifier thus:
$J = \tfrac{2}{5}.m(\Delta.r)^2$
the radial modifier (Δ) defines a body's radial centre of mass according to its variable density.

If you know a body's angular velocity (ω) you can calculate its radial modifier (Δ). Alternatively, if you know its radial modifier, you can calculate its angular velocity.

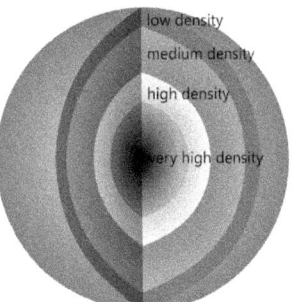

Rocky Planet

When used in conjunction with Core Pressure, 'Δ' can help us to determine a body's internal structure.

Fig 19

7.2.2 Earth's Core

The earth's core is a large ball of [mostly] iron, the spin of which is controlled by the potential energy of the earth's force-centre (its sun).

Spin in the earth's mantle, however, is dominated by the kinetic and potential energies of its moon. Both of these energies, which are driving spin in opposite directions, are creating internal friction between its core and its mantle.

Rotation of the earth's core ($\approx$7E-05 radians per second relative to its mantle) generates its magnetic field and the friction that is the source of the earth's internal heat.

Internal frictional heat from planetary spin is the source of all universal (electro-magnetic) energy, including the stars.

Here's an interesting question! *What would happen to the earth if it acquired another significant moon?*

7.2.3 Earth's Magnetic Field

The earth's lunar tilt angle (Fig 20: α = 23.4°) is a clear indication that the earth acquired its moon from outside its solar system (galactic comets), as is the case for all of the moons in our solar system, and perhaps, most of its planets.

Spin in the earth's core is dominated by our sun and spin in its mantle is dominated by its moon. The resultant relative spin rate, together with the magnetic properties of what we understand today as *mass*, in the earth's iron-rich core and mantle are responsible for generating the earth's polar magnetic field.

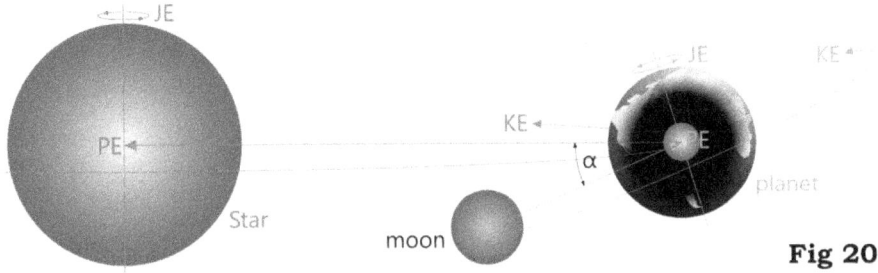

Fig 20

If a planet's lunar orbital plane is not coincident with its solar orbital plane, its mantle and core will spin on different axes, generating an angular difference between its true (physical) North and its magnetic North (Fig 21: β = 6.277°)

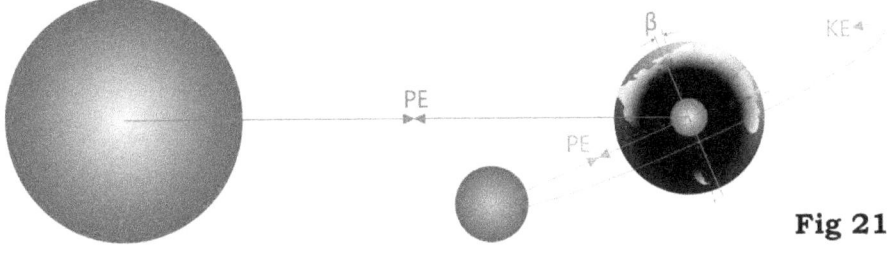

Fig 21

This phenomenon is confirmed by; the lack of a magnetic field in Venus and Mercury, and the massive magnetic field in Jupiter that acts at 90° to the orbital plane of its moons, similar to the earth. In fact, all planets and stars with satellites will generate a magnetic field ≈normal to the orbital plane of their satellites.

7.2.4 Magnetic Reversal

A worrying aspect of the earth's magnetic field is that the relative spin induced in the earth and its core by our moon and our sun will not reverse unless either the earth or its moon changes orbital direction, which is highly unlikely. Something external to the earth (extra-terrestrial) must therefore cause this reversal to occur periodically.

It would appear that the earth's magnetic reversal can only be explained by flipping it through 180°; switching north and south poles!

We already know that our solar system has a number of orbiting comets, so it is highly likely that the Milky Way also has its own 'comets', and these could be planet sized. Therefore, a large galactic comet may well be responsible for flipping the earth and/or any other planet in the solar system as it passes close by.

7.2.5 No Moon!

Using this calculation method, it has been possible to determine what would happen to the earth's spin if it lost its moon.

If the earth lost its moon, it would also lose its tilt (and its seasons), there would be insufficient internal heat energy to drive its continental plates and its magnetic field would vanish. The loss of internal friction would cause the earth's surface temperature to fall by about 200K.

The earth would therefore behave similarly to Venus except for its atmosphere, most of which would liquefy/solidify because the earth' surface receives only 52% of the sun's radiated heat compared to Venus.

Without its moon, one earth day would be 12450.1516 hours (<519 current earth days) and the sun would rise in the West and set in the East just like Venus. There would be no seasons because the earth would lose its tilt.

7.2.6 Chicken & Egg

Which came first; spin or orbit?

What you see in most films and documentaries is that the sun starts spinning and the planets follow it around. This is of course 'back-to-front'.

In order to generate spin, you need an appropriate energy. Spin theory teaches us that if a sun, planet or moon sat alone in space it would not spin.

Spin in our sun was first induced by the rotational energy in its force-centre (Hades) and it would have continued to spin at this rate had it not acquired its satellites (planets). However, our sun actually rotates at more than ten times this speed.

If angular kinetic energy in a force-centre induced orbital kinetic energy in its satellites, this transfer of energy would slow down the force-centre's rotation, which is obviously not the case. I.e. kinetic energy in the planets can and does induce rotational kinetic energy in the sun.

Therefore, the planets must have been orbiting long before the sun achieved rotation anywhere near its current rate.

The same argument applies to a spiral galaxy. Our sun got its initial spin (<2E-07 radians per second) from the spin energy in Hades, but Hades is in a linear orbit, so it generates little internal frictional heat through spin. Therefore, all of Hades spin comes from its orbiting satellites.

So, orbits came first!

7.3 Core-Pressure

This astro-physical property appears to have been overlooked to date but can be readily defined using Newton's revised formula: $p = G.m_1.m_2/A^2$ in which m_1 is the mass inside radius 'r' and m_2 is the mass outside 'r' and 'A' is the spherical surface area at radius 'r'.

An example calculation that yields a surprising, but understandable result has been performed for the earth.

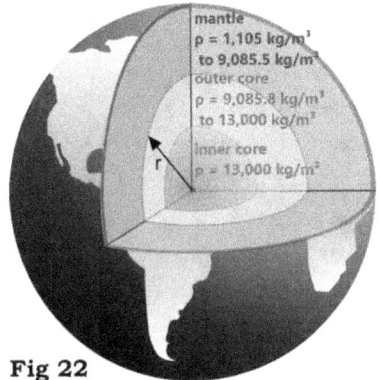

Fig 22

Most publicly available sources claim that the earth has a core density of somewhere around 13000 kg/m³, which, given the incompressibility of iron, appears unlikely unless it contains a much larger percentage of heavy metals than is currently believed. However, whilst the structure described in Fig 22 provides the correct total mass and polar moment of inertia for the earth, because of its relatively low percentage mass, reducing the core density will have little effect on the density of the upper mantle material (immediately below the earth's crust) that cannot be much more than that of water.

Whilst this discovery may be unexpected, it is perfectly logical. It is currently claimed that mountain roots beneath the earth's crust eventually fall into the upper mantle, causing the crust to rise locally and temporarily. This event would be difficult to explain if the density of the mantle were greater than, or even close to, that of the crust. It is not so difficult to understand, however, if the upper mantle is a hot, gaseous, pressurized, cauldron of matter with a lower density than the crust it is supporting. Moreover, it is also easier to see where all the volcanic activity comes from. The same applies to subduction zones where crust material is pushed into the upper mantle. The sinking of the earth's denser, cooler continental crust into its mantle is actually aided by the relatively low density in the upper mantle and also generates the mantle plumes as it sinks towards the earth's core where it heats up and rises to the surface.

7.3.1 The Structure of Celestial Bodies

All the celestial bodies wandering the universe - such as comets, moons, planets and stars - are composed of the same matter that originally comprised the ultimate-body immediately prior to the last 'Big-Bang'. They therefore all have the same age.

When cold (such as galactic force-centres) these celestial bodies comprise the same matter that was created during the previous universal period and held in the ultimate-body. All galactic satellites that have collected satellites of their own, will generate internal heat. These are the stars and planets.

Due to the combined mass of their galactic force-centre and secondary satellites (planets), galactic satellites (stars) may eventually generate sufficient internal heat to create neutrons and fissionable energy, the by-product of which is hydrogen (H), at which point they become bright. The core of a star is mostly iron (the commonest and largest stable element), its outer mantle is a store of neutrons and its outer surface comprises proton-electron pairs in the form of hydrogen and helium atoms, which is the reason they can emit electro-magnetic energy.

The fusion generated in galactic force-centres produces no heat (energy), which is why they are cold. Fusion is an instantaneous event, once achieved, any energy released will cease, unless the mass of the body continually increases, which is not the case.

Neutrons are the stores of energy that will be contained within the ultimate body at the end of the current universal period and provide the explosive energy for the next 'Big-Bang'.

The *largest* - and outermost - secondary satellites (planets) normally collect sufficient satellite populations (moons) of their own to cause their outer crusts to melt or 'skin'. These planets will become gas planets.

The term 'skin' means an active [hot] surface covered by skin that forms due to the cooling effect of atmospheric gases.

All *active* celestial bodies (those generating internal heat) will comprise matter of greatest density at their core gradually reducing with radial distance from it. Their structure is defined by their radial modifier (Δ)

8 A Universal Theory (how it works)

Today's universal hypotheses, that 'solar systems accrete from gas in space', 'galactic force-centres are black-holes' and 'singularities generated the '*Big-Bang*', make no mathematical or logical sense and can be easily discredited. There is, however, a theory that does make mathematical and logical sense and is extremely tricky to disprove:

All universal energy is held in neutrons, which are created through fissionable decay in bright stars.

The relative rotation inside a satellite that is also a force-centre, generates its internal heat through friction.

Fissionable decay will occur in the core of a force-centre if its own mass, and that of its satellite population, is sufficient for spin to generate, and hold, heat at the neutronic temperature. The masses required for such an event, means that it is most likely to occur in galactic satellites (stars).

This same process (spin) is also responsible for generating the internal heat within planets. Whilst planets are unlikely to collect the satellite mass necessary for fission energy generation, the outer planets in any solar system are free to collect sufficient satellite mass to generate the internal heat required to melt their surface crusts; these are the gas planets.

All galaxies are travelling in linear orbits away from the latest '*Big-Bang*', but slowing down due to the magnetic (gravitational) attraction from the great attractor. Outward travel will eventually cease and reaccrete into another ultimate body, when a new '*Big-Bang*' will occur.

The temperature of outer space ($\approx 2.7255K$) is the electro-magnetic energy radiated by all the active stars and planets in the universe.

During any universal period, new elements are created either via:
1) fission in stars (hot), or
2) fusion in galactic force-centres, the great attractor and the ultimate body (cold)

When the core-pressure within the ultimate body is sufficient to exceed the coupling ratio (φ), two adjacent neutrons will touch, causing each to split into its component parts (a proton and an electron), generating an explosive chain reaction (*Big-Bang*). This is the same process that occurs within nuclear reactors and atom-bombs#.

in which neutron energy is released by artificially increasing the neutronic ratio of all the star's core elements to more than 1.5

All the matter contained within the ultimate body is, consequently, ejected outwards at a velocity of approximately 1.777E+07 m/s due to the release of about 7.4E+60 Joules of kinetic energy. Conversely, the potential energy generated by the residual matter left behind after the last '*Big-Bang*' (the great attractor; ≈1.04E+46 kg) (gravity), is gradually slowing down this outward travel. Today it is claimed (NASA) that the velocity of our Milky Way is about 2.3E+05 m/s. Eventually, this outward travel will cease altogether and all universal matter will return to reunite as another ultimate body, after which, it will explode again. All the neutrons lost during each 'Big-Bang' are recreated in the bright stars during each subsequent universal period.

Long ago, life was accidently created on one planet. After countless 'Big-Bangs' its proteins are now distributed throughout the universe.
During any universal period, these proteins will have the opportunity to evolve into complex lifeforms on planets that generate just enough internal heat (spin) to retain a thin surface crust and also receive sufficient stellar radiation for photosynthesis. In any solar system, these planets are most likely to exist between the moonless planets closest to their [force-centre] star and the gas planets furthest from it.

All the matter in the universe is constantly being recycled during all universal periods. This repetition can occur indefinitely with no outside help and obeys the single-most important scientific law; 'the conservation of energy'. Moreover, it requires no myth or abstract theory to explain it.

But it works; perfectly!

The Universe

The universe is in fact, an enormous energy generator that uses magnetic & electrical charge to create electro-magnetic energy (heat and light); and it works like this:

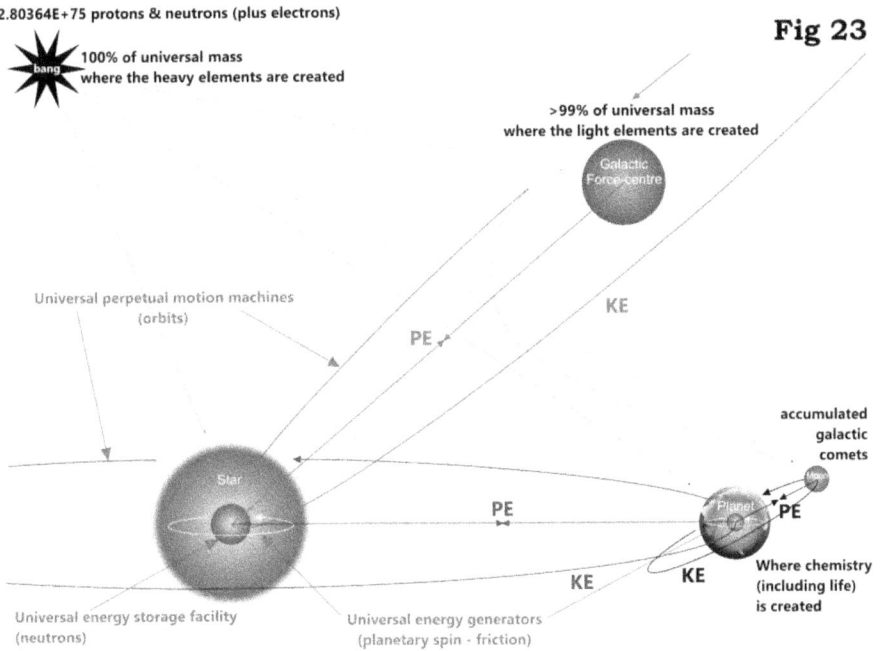

Fig 23

8.1 The Ultimate Body

This term is used here to describe a spherical body that contains all universal matter (>2.80211538385852E+75 proton-electron pairs), the mass of which (>4.68943137883966E+48 kg) is necessary to overcome the coupling ratio (4.40742111792334E-40) in the elements at its core and thereby initiate a nuclear chain reaction; what we call a *'Big-Bang'*.

The coupling ratio is the relationship between the electrical [charge] repulsion between two adjacent protons, as described by Coulomb, and the magnetic (gravitational) force as described by Newton.

When this mass is combined in a single, perfect sphere, the pressure generated at its core will be sufficient to force two neutrons in adjacent atoms to touch, causing them to split into their component parts; an alpha particle (a proton) and a beta particle (an electron). The kinetic energy in the protons released is sufficient to split another adjacent neutron, etc., etc., etc.; a chain-reaction. This chain-reaction is what we all know as a 'Big-Bang', which releases ≈7.4E+60 Joules of energy, breaking the spherical mass into pieces that are ejected into empty space at a velocity of ≈1.8E+06 metres per second.

During the build-up of universal matter towards the end of a universal period, the core pressure within most of this mass is sufficient to fuse small atoms into larger atoms. The largest elements will be fused at the body's core, reducing in size towards its surface. More than 80% of the ultimate body's mass will be capable of fusing the smallest elements at ≈3K; the temperature of outer-space.

The actual mass in our universe will be greater than the aforementioned mass, because the ultimate body is unlikely to be a perfect sphere.

The Great Attractor is the matter left behind after a *'Big-Bang'*. Its mass is approximately 1.04E+46 kg (refer to Chapter 8.2.1) and responsible for slowing down outward travel of the galaxies that will eventually reunite into another *'Big-Bang'* at the end of the universal period.

8.2 The Universe

The universe is simply the empty space that contains all the matter ejected from the ultimate body following a *'Big-Bang'*. It is ellipsoid in shape, and all the galaxies it contains are travelling outwards at roughly the same velocity.

Whilst the kinetic energy in all this ejected matter was sufficient to achieve an initial velocity of ≈1.77E+06 m/s, the potential energy (magnetic charge or gravity) in the Great Attractor is slowing down its outward travel. Eventually, it will cease moving outwards and return to a common point, where it will once again reunite as another ultimate body (Fig. 24).

The most massive celestial bodies ejected during a *'Big-Bang'* will become galactic force-centres as they gather together the nearest smaller bodies, the slightly different velocities of which will result in their orbital trajectories (galaxies).

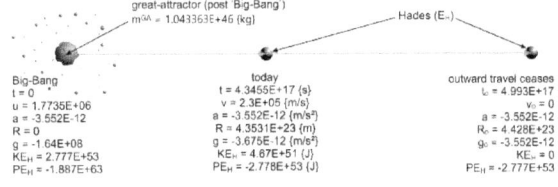

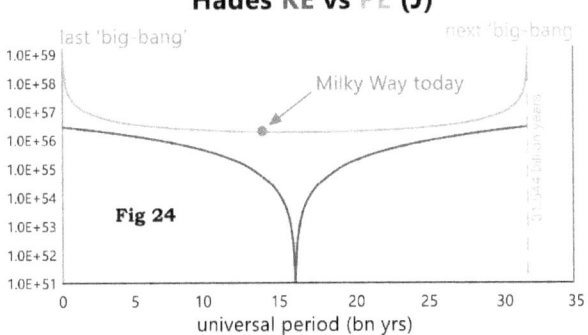

Fig 24

The temperature of universal space; ≈2.7255K, is the electro-magnetic energy radiated by all the active stars and planets in the universe.

The reason we know it is not heat left over from the last *'Big-Bang'* is because heat is electro-magnetic energy, which travels at ≈3E+08 metres per second, i.e. faster than the universe is expanding. So, any heat left over from the last *'Big-Bang'* would be well outside the universe by now.

All the matter in the universe, from galactic force-centres to comets originated from the same ultimate body, at the same time and comprise the same basic matter. If the last *'Big-Bang'* was 13.6 billion years ago, then that is the age of all the matter in it; including the earth and our sun.

The space between all celestial bodies is empty of all matter; it is a perfect vacuum.

8.2.1 Universal Size and Age

NASA predicts the current age of the universe as 13.77bn years and that the Milky-Way's current velocity is 2.3E+05 m/s. If these figures are correct;
universal age today: t = 4.34548152E+17 s
velocity of Milky Way today: v_1 = 230,000 m/s
ultimate body mass: 4.68688E+48 < m_u ≤ 1E+49 kg
neutrons in m_u: N_n = 1.49675E+75
energy per neutron: E_n = 1.637856E-13 J
energy released during the last big-bang: KE > 7.35440E+60 J
based upon the splitting of 3% of the ultimate body's neutrons (Little-Boy)

The size and age of the universe may be estimated as follows:
Initial velocity of universal matter: v_o = √[2.KE/m_u] = 1.77349841E+06 m/s
gravitational acceleration today: a = (v-u)/t^2 = -3.5519617407994E-12 m/s^2
distance travelled since '*Big-Bang*': R = ½.(v^2+u^2)/a = 4.353082659E+23 m
great attractor mass: m = PE.R / G.m = 1.04336261407223E+46 kg
outward movement will cease at: R_o = u^2 / 2.a = 4.42754855113119E+23 m
and:
t_o = u/a = 4.9931E+17 s (15.822 billion years) after the last '*Big-Bang*'

All universal matter will then return over the same period (t_o), when it will reaccrete into an ultimate body and thereby compromise its innermost neutrons, resulting in the next '*Big-Bang*'.

This all means that *if* NASA's data is correct:
the total mass of the universe, including the 'Great Attractor' is;
4.68688E+48 < m_u ≤ 1E+49 kg
and we are 43.5% through our current universal period; i.e. it has another 17.87379 billion years yet to run;
and because EME travels at 'light-speed', any heat radiated at the time of the last '*Big-Bang*' would have travelled 1.30274E+26 metres by now; i.e. it would be outside the universe (≈4.3531E+23 m). Therefore, the temperature of outer space cannot be left over from the last '*Big-Bang*'. It is actually the heat radiated by all the stars (and planets) in the universe.

Moreover, the time (t) for light to travel across from the other side of the universe (2.904064E+15 s) is less than its age (4.34548152E+17 s). So, a telescope with the necessary resolution, should enable us to see the entire universe from Earth today; but not at the same time!

8.3 Planetary Spin Energy

Planetary spin is responsible for generating the internal frictional heat and protective magnetic field in all celestial bodies that are both satellites and force-centres. Most of a planet's surface heat is always derived from its internal friction rather than that radiated by its star.

The kinetic energy in orbital satellites causes its force-centre matter to rotate. The potential energy between a satellite and its force-centre prevents a satellite's core from rotating. Therefore, the relative rotation between the core and mantle matter within any satellite that is also a force-centre will generate internal frictional heat.

Whilst force-centres that are not also satellites in elliptical orbits (e.g. galactic force-centres) and satellites that are not also force-centres (e.g. moons) will spin, they generate negligible internal [frictional] heat; most of their heat is from external radiation.

8.4 Galaxies

Galaxies comprise a [single] galactic force-centre; its satellites (stars and comets), its sub satellites (planets and comets) and its sub-sub-satellites (moons).

All galactic force-centres and satellites comprise similar matter as all other celestial bodies.

Galactic force-centres are in a linear orbit, travelling in a straight line - at similar velocities - away from the great attractor. Because the great attractor does not spin, galactic force-centres will generate negligible internal frictional heat, and are therefore cold; and dark.

All galactic satellites are either; stars (bright or dark) or [galactic] comets.

Galactic comets tend to be small, relative to stars and usually follow elongated orbital paths. Just as occurs to all orbits, as a satellite passes close by and is attracted to, another satellite, its orbital axes will rotate (about its force-centre). This will cause it either to impact, or bring it into the orbital grasp of a celestial body with superior magnetic (gravitational) attraction. Galactic comets are the source of all galactic sub-satellites (planets) and sub-sub-satellites (moons).

8.5 Solar Systems

An orbital system is a single force-centre together with its satellites (planets & comets) and sub-satellites (moons).

'Solar' implies a force-centre that has accumulated sufficient planetary mass to generate the internal [frictional] heat (through spin) required to achieve the neutronic temperature (623316124.717178 K) in its core elements; a bright celestial body.

We today refer to a solar system as a galactic sub-orbital system with a bright celestial body as its force-centre; a star. However, if Jupiter acquired sufficient additional satellites to generate the neutronic temperature in its core elements, it and its moons, would also be a '*solar system*', <u>and</u> a 'planet'.

A galactic satellite that has accumulated insufficient planetary mass to generate the neutronic temperature in its core elements, is also part of a galactic orbital system but not a *solar* system, because it is not bright.

A galactic satellite that has accumulated no satellites of its own will be dark (cold), because it will have no means of generating internal [frictional heat].

It is expected that most galactic satellites are dark.

8.6   Stars

Just as with all force-centres that are also satellites, stars (galactic satellite) generate internal frictional heat from spin. Stars may be dark, dim or bright.

Dark stars are galactic satellites that have accumulated little or no satellite mass. They are therefore cold; they radiate little or no electro-magnetic energy. But that which they do radiate is extremely low frequency and difficult to detect. These stars are invisible to our eyes in the night sky.

Dim stars are galactic satellites that have accumulated sufficient satellite mass to generate internal frictional heat through spin, but insufficient to achieve the neutronic temperature. These stars resemble large galactic planets. Whilst they radiate higher frequency electro-magnetic energy than dark stars, they remain invisible to our eyes in the night sky.

Bright stars are galactic satellites that have accumulated sufficient satellite mass to generate the internal heat (through spin) to achieve the neutronic temperature in its core elements. This process creates neutrons and therefore fissionable decay. The additional [fissionable] energy adds considerable intensity to the heat (and light) that these stars generate (and radiate). These are the stars we see in the night sky. The neutrons created in bright stars hold all universal energy and are responsible for the energy released during each '*Big-Bang*'.

Stars are far too hot, and they have insufficient mass, to fuse elements together, and no two of their neutrons can touch through natural means. Therefore, no star can explode of its own accord. Moreover, all of a star's elements, from hydrogen to uranium - the heaviest of which (uranium) will be in the star's core and the lightest (helium and hydrogen) will be at its surface - are in gaseous form.

However, when two sufficiently massive stars collide, neutron contact is possible, creating a nuclear chain reaction; a mini '*Big-Bang*', ejecting their gaseous matter into space in the form of a gas cloud. Because a star's core matter is essentially gaseous, its radiated EME will be of all wavelengths, from gamma to radio, creating the multi-coloured clouds we see in the universe. Such an event is what we currently call a nova. Eventually, all these atoms will be attracted through magnetism (gravity) to the celestial bodies within and close to the cloud. But it will take many millions of years to occur due to the distances involved.

8.7 Planets

Planets are stellar satellites. The proximity of a planet to its force-centre (star) will influence its ability to trap galactic comets (moons).

It will be very difficult for planets close to their force-centre to trap galactic comets due to the much stronger magnetic (gravitational) attraction of their star. Therefore, the planets nearest to their star will be hot at their surface due to high stellar radiation, but their cores will be inactive and relatively cold. Moreover, these planets will generate no protective magnetic field.

It is easy for planets furthest from their force-centre to attract galactic satellites due to the lower magnetic (gravitational) influence of their stellar force-centre. Whilst these planets receive negligible stellar radiation owing to their proximity, they will generate much greater internal frictional heat due to their superior satellite (lunar) population, sufficient in-fact, to melt their crusts. This is where we will find most of a solar system's gas planets.

Planets between these two orbital regions may be able to trap one or two moons, giving them the opportunity to generate sufficient internal frictional heat to keep the planet warm, but not enough to melt their crusts. These planets are also likely to be sufficiently close to their star to receive the stellar radiation for photosynthesis to occur on their surfaces. These planets are capable of creating life. Whilst the conditions achieved here on Earth will be rare, there will be other planets out there in our own and other galaxies that will allow life to evolve.

8.8 Moons

Moons are simply planetary satellites. Because they are unlikely to attract satellites of their own, they will generate no internal heat through planetary spin.

Therefore, a lunar satellite will simply orbit its planetary force-centre, always presenting to it the same face. The only heat it will receive is that radiated by its planet and its star according to its proximity to either source.

A lunar satellite will become active (warm) if it is orbiting sufficiently close to a particularly large planet. The differing potential energies imposed on its internal core between its apogee and its perigee will generate heat by altering its shape. Whilst this is not the same heat generation process as planetary spin, the heat generated in both cases is due to internal friction.

It may not be obvious at first, but moons are essential for the creation of life, which only occurs on planets. And most of a planet's heat is generated by its lunar satellites. Without moons, there would be no universal life.

8.9 Our Galaxy: The Milky Way

A detailed analysis of our solar system reveals the correct amount of matter together with all the velocities and orbital shapes with no inconsistencies. This is how we know that Isaac Newton's theories are correct. Contrary to popular belief, the same can also be said of our Milky Way galaxy.

According to a scientist in the 1930s, Newton's laws predicted that the total mass in our galaxy (the Milky Way) could not be accounted for from observations and suggested that the entire Milky Way must be full of dark matter (which comprised sub-atomic particles) to prevent the spiral arms of the galaxy being ejected under centrifugal force.

In fact, this claim has now escalated to the extent that it has been embraced by 99% of the world's astrophysicists who believe that 85% of the entire universe comprises dark matter in the form of sub-atomic particles, and a great deal of time, money and effort has been spent looking for it.

It is difficult to understand this situation given that assuming NASA's data for our sun's perigee distance, its orbital period and the estimated [number of] solar masses in the Milky Way are all correct, the sun's orbit within the Milky Way can be fully resolved using Newton's laws of motion and planetary spin theory. Even if NASA's data is incorrect, the same calculation can be carried out using alternative input data, revealing slightly different results, but no less successfully. There simply isn't any need for dark matter.

Apart from which, a gas cloud cannot act as an orbital force-centre. Moreover, why do these same scientists also believe in *'black-holes'*; they cannot both be necessary!

8.9.1 Hades

Hades is a name I have adopted for the Milky Way's force-centre for easier reference.

We know it exists because it is a fundamental law of nature that every orbital system *must* have a force-centre.

So why can't we see it?

Hades is not an elliptical satellite, therefore it cannot generate internal heat energy (through friction). It is therefore cold; it emits almost no electro-magnetic radiation.

Moreover, it has a diameter of ≈1.563E+12m,
whilst our Milky Way has a half-major axis of ≈2.5E+20m
the ratio being; 1.5355E+08
and the radius of an iron atom's outermost proton-electron pair = 7.6E-08m

$\quad$ 1.5355E+08 x 7.611353E-08 = 11.6873m

So, it would be like looking for a black atom in the centre of an 11.7 metre diameter disc of black-iron.
That's why you can't see it.

But it *is* there, and it is the primary source of the heat energy generated in all its orbiting stars.

At the time Relativity was theorised, nobody else was aware of the exigency of force-centres in *every* orbital system or of [planetary] spin theory. The scientists concerned therefore misunderstood the effect of galactic population on orbital shapes; hence the misguided invention of dark matter.

Moreover, if these scientists had known of Hades and the laws of station-keeping, they would have realised that the deformation of space-time could not work.

8.9.2 Our Sun

Δ = 0.318284697814735   ρ = 1409.782932 kg/m³

As with all stars, our sun originally comprised the same elemental matter as all the other celestial bodies in the universe. However, its massive force-centre (Hades) and significant satellite population (our solar system) is sufficient to generate the internal heat required to convert elemental proton-electron pairs into neutrons, the by-product of which is hydrogen (H) and helium (He).

It is currently claimed that our sun is creating elements from hydrogen (H⁺) through fusion and that it is [apparently] growing in size with age.
The problem with this scenario is that fusion *increases density* and therefore *reduces size*. Moreover, why is Hades cold if fusion generates heat, as Hades is far more likely to generate fusion than our sun.

Its hot hydrogen blanket is a star's atmosphere, which generates the electro-magnetic energy we receive as light and heat. The only reason we can see it is because the hydrogen and helium at its surface comprises proton-electron pairs, which *are* capable of emitting electro-magnetic radiation. Our sun's atmosphere would be invisible (and cold) if its surface was lone protons (H⁺).

8.9.3 Mercury

$\Delta = 0.81286219642311$ **$\rho = 5427.012135$ kg/m³**

Mercury's 'Δ' value and its density show it to be an iron-rich planet with an internal structure much more evenly distributed than is the case for Venus or the earth.

Because it has no moon to generate internal friction it cannot have a particularly active core and is therefore unable to generate any internal friction or volcanic activity, hence no surface gases (atmosphere). This is also the reason why the surface of the planet's *'far-side'* is so cold, despite being so close to the sun; it has no internally generated heat.

Mercury's matter is similar to when it was ejected during the *'Big-Bang'*.

Mercury is therefore a solid piece of iron-rich matter with a cold core. There will be little or no volcanic activity on Mercury as it has no tectonic plates and negligible internal heat.

There is little to distinguish it from, say, Pluto other than its composition and the temperature of half its surface, and that Mercury has acquired no moons of its own because it is too close to the sun. Any passing galactic comets will have been trapped by the sun's greater mass and proximity.

8.9.4 Venus

Δ = 0.681231909980155   ρ = 5242.664311 kg/m³

Venus's 'Δ' value and its density show it to be an iron-rich planet with an internal structure more evenly distributed than is the case for our earth but less than that within Mercury. Its origin is similar to that of Mercury (the ultimate-body).

It spins the opposite way to the other planets because: a) it has no moon(s) to drive it in the other direction, and; b) owing to its size, the sun's angular kinetic energy dominates Venus's spin.

Whilst it has no moon, its much greater mass (than Mercury) is sufficient to enable the sun's spin energy to induce some internal friction (heat) …

$E_2 = E_1 - E_0 \quad (E_3 = 0)$

… giving it the capacity to generate some, albeit minimal, volcanic activity, and therefore surface gases (atmosphere).

Venus's internal matter is therefore more likely to gravitate towards its core than within Mercury. This also means that whilst there will be insufficient heat to generate tectonic plates (Venus's crust is too thick), there should be sufficient to generate sporadic (and random) volcanic activity, which together with its proximity to the sun, will generate more than enough heat to vaporise its atmospheric water.

The reason Venus's surface material is so flat is because its volcanic activity is far less aggressive than on the earth due to its significantly lower internal heat and much thicker crust.

Venus is, and always has been, too close to the sun to allow water to exist on its surface in liquid form. The mass of water vapour maintains the surface temperature of Venus.

8.9.5 The Earth

$\Delta = 0.3342776982996 \qquad \rho = 5506.351327 \text{ kg/m}^3$

Unlike Venus, the earth has a substantial satellite driving its angular motion, contrary to which, the sun is trying to drive the earth's core in the opposite direction and slow it down. Therefore, the earth's mantle and its core are revolving at different rates, generating internal friction and, as a result, internal heat.

This relative rotation (7E-05 °/s) is sufficient to generate the internal heat through friction that drives the mantle plumes, and in turn drives the tectonic plates. The earth gets its earthquakes, volcanoes, weather and its carbon-cycle from this internal activity. The moon's tilted orbital plane gives the earth its seasons. Along with the earth's proximity to its sun, all this internal and surface activity provides an ideal environment for life to flourish.

Spin and core-pressure analysis has shown us that the density of the earth's mantle material just below its crust is not much more than that of liquid water, allowing heavier material, such as mountain roots to fall from the crust.

Referring back to earlier attempts to age the earth via heat loss:

Kelvin's assessment of the age of the earth through heat loss may not have been wide of the mark if the earth's internal heat is indeed left over from its birth. But all this tells us; is that if his hypothesis was correct, the earth would have lost all of its internal heat within the first 20 to 100 million years of its life, which is clearly not the case.

Rutherford's subsequent claim that additional heat sources would have increased the earth's perceived age was, however, wide of the mark, as there is no way the heat generated by the earth's radioactive substances could account for the heat it currently possesses.

The magnitude of the earth's internal heat means that it must be constantly generated, and its only source can be from friction between its inner core and its mantle matter. This heat is the source of the earth's mantle plumes. It is greatest at the core-mantle interface and rises to cool below the earth's crust.

The Universe

Less than 0.04% of earth's atmosphere is generated by its mantle activity, all the rest (N_2, O_2 & Ar) has been created by its surface life and potassium decay. Less than 40% of the earth's atmospheric heat is generated by the sun's radiated energy. The rest (>60%) is generated by its mantle heat.

The early atmosphere must have been very thin indeed (almost non-existent): mainly sulphurous chemicals, hydrogen and carbon dioxide, but with only a tiny fraction of today's density and volume (<0.039%). Therefore, meteorites would have been considerably colder at the time of impact than would be the case today. Only friction generates heat, not compression and rupture. Given this and the following:

a) The earth's internal heat is generated by internal friction.
b) Before acquiring a moon, the earth was cold.
c) The earth's surface water was liquid well before 4bn years ago.
d) Impacting masses create very little heat if they don't have to pass through a heavy atmosphere (no friction).

The earliest period of earth's existence must have been quite cool. It will have had no moon and therefore no tectonic plates or volcanoes. Its first volcanoes, after acquiring its moon, would have been distributed at random. More than 3.8bn years ago there were pebble beaches on the surface of the earth (The Isua Greenstone Belt, Greenland) indicating that it must have had active surface water at least 4bn years ago. It is highly likely therefore, that the images we see of a hot earth during creation are incorrect as there was very little to generate the heat and there appears to be no supporting geological evidence. Moreover, the earth was never born, like all the other celestial bodies in the universe; it was ejected from the ultimate-body 13.6bn years ago. Moreover, the earth's internal heat has been possible only since it trapped its moon, and we have no idea when that was.

The meteorites we collect that have been aged at 4.66bn years old are nothing more than left-over rubble from the destruction of the planet that once orbited at the asteroid belt.

8.9.6 Mars

Δ = 0.0023170868178197   **ρ = 3934.080869 kg/m³**

It is assumed that the information sent back from the various planetary missions to Mars have confirmed its mass.

If Mars is an iron-rich planet, its density and 'Δ' value would tend to indicate that the planet must be hollow.

It is probable that the energy induced within Mars by its second lunar acquisition was sufficient to blast its iron-rich mantle material onto its surface, leaving its internals full of voids, which would explain its apparent density and its 'Δ' value. All of its surface water would no doubt have migrated into its voids and is occasionally released only under heavy meteorite impact.

If correct, this would explain the planet's low density, despite being an iron-rich planet.

During this early activity, sufficient internal heat would have been generated to maintain Mars's surface water in liquid form despite its distance from the sun and may also have been able to generate life earlier than the earth. Mars's red surface colour (rust), along with the presence of surface canals, may well indicate that it had accommodated oxygen-emitting plant life along with liquid surface water before its mantle was blown out.

Apart from the earth, Mars is by far the most interesting planet in our solar system, because *if* the following are true:
1) it is a hollow iron planet,
2) it had liquid water on its surface,
3) it has hosted oxygen-emitting plant-life,
4) its water has found its way into the planet's interior,
… it is possible that the kinetic energy in Phobos is keeping Mars's internal water liquid, which would mean that:
a) it may contain an atmosphere (of some description) internally,
b) it may contain life internally, and
c) it may be possible to access its internal voids.

Could it be that there is an atmosphere and water inside Mars, being kept liquid by an over-active moon?

8.9.7 The Asteroid Belt

$\Delta \approx 1$ (est.) $\rho = 2090$ kg/m³

It appears that there was once a planet at position '5' in our solar system and that it was impacted by a galactic comet, and is probably the source of many of our meteorites today and in the past. Perhaps this impact created the comet that hit the earth 63mn years ago; i.e. the asteroid belt was created 63mn years ago.

One of the reasons we know that our sun and its planets did not accrete from rocky particles is that the Asteroid belt remains a loose collection of rocks. I.e. planets do not accrete through gravity.

Because the moons in our solar system are trapped from galactic travel, it would appear from the Asteroid incident, that galactic comets can be quite substantial. However, whilst our own comets tend to orbit in cycles of hundreds of years, galactic comets will orbit in millions of years. But keep your eyes open!

8.9.8 Jupiter

Δ = 0.0227806696137   ρ = 1326.216812 kg/m³

Jupiter comprises the same matter and has the same age as every other planet, star, comet, etc., in the universe, i.e. that of the current universal period.

Jupiter is a gas planet simply because its *mass* and orbital distance have together enabled it to trap many sizeable moons; Jupiter's largest moon (Ganymede) is approximately twice the mass of the earth's moon. Together with the mass of its force-centre, its satellite population (>50) is sufficient to generate the heat necessary to melt its crust, but not sufficient to generate neutrons (or fission). This is the reason for its apparently low density. Its average body density should be calculated on the basis of an outer diameter that excludes its gas cloud and is likely to be similar to that of the solar system's inner planets.

Whilst its surface may be molten, its heavy surface gases will probably extract sufficient heat to form a protective skin over its surface, but it will be relatively thin.

Jupiter's weather is created by its moons orbiting in both directions (prograde and retrograde) that together generate sufficient competing kinetic energies in the planet's surface gases to account for its violent weather. The Red-Spot rotates because of its two adjacent gaseous layers rotating in opposite directions, which is caused by the opposing orbital directions of its moons.

If Jupiter's density is similar to that of the inner planets (5392 kg/m³); its body diameter should be 87605252 m (6.875 times that of the earth), giving it a gas-cloud cover thickness of 26107374 m

Using this information, the body of Jupiter is spinning at 1.9344E-05 radians per second and its properties are:

Δ = 0.33   ρ = 5392 kg/m³

8.9.9 Saturn

Δ = 0.0140600109265   **ρ = 687.1230137 kg/m³**

Just as for all the universal bodies, Saturn comprises the same matter and is of the same age as every other planet, star, comet, etc., i.e. that of the current universal period.

Whilst Saturn is only 30% of Jupiter's mass, its largest moon (Titan) is almost 92% as massive as Jupiter's largest moon, therefore Saturn will be much more active [internally] than Jupiter.

This is no doubt the reason for its extremely low density. A significant percentage of its body matter has been converted to gas and is likely to generate more active weather than Jupiter.

Saturn's rings are most probably the remains of a satellite that was pulled apart by the huge gravitational forces induced by orbiting so close to such a large planet. If the satellite was an ice moon, it would not have required much kinetic and potential energy to pull it apart.

Saturn and Jupiter have collected the most moons because they are the most massive planetary satellites in our solar system.

If Saturn's density is similar to that of the inner planets (5392 kg/m³); its body diameter should be 58607472 m (4.6 times that of the earth), giving it a gas-cloud cover thickness of 28926264 m

Using this information, the body of Saturn is spinning at 1.6379E-05 radians per second and its properties are:

Δ = 0.28   **ρ = 5392 kg/m³**

8.9.10 Uranus

$\Delta = 0.0249376193237$ $\qquad$ $\rho = 1270.415139 \text{ kg/m}^3$

Uranus is another 'gas' planet just like Jupiter and Saturn for exactly the same reasons, but being so much further away from its force-centre's radiated heat, much of Uranus's surface gases will remain in liquid form, hence its relatively (to Saturn) high density.

Whilst Uranus has at least 30 moons, all but the smallest, which are also the furthest away, orbit in the same direction. It is not therefore expected that the same level of disruption will be seen on the surface of Uranus as that seen on Jupiter and Saturn.

If Uranus's density is similar to that of the inner planets (5392 kg/m³); its body diameter should be 31328902 m (2.46 times that of the earth), giving it a gas-cloud cover thickness of 9697549 m

Using this information, the body of Uranus is spinning at -1.5186E-05 radians per second and its properties are:

$\Delta = 0.27$ $\qquad$ $\rho = 5392 \text{ kg/m}^3$

8.9.11 Neptune

Δ = 0.06523792740988   ρ = 1637.934377 kg/m³

Neptune is also a gas planet, but so far from its sun that much of its surface gases will remain in liquid and even solid (ice) form, hence its greater (than Uranus) density. That said; it still has a satellite population sufficient to melt its crust.

Whilst Neptune has only 15 or so moons, its largest moon (Triton) orbits in the opposite direction to most of the others. It is expected therefore that Neptune's surface matter will suffer significant levels of disruption.

If Neptune's density is similar to that of the inner planets (5392 kg/m³); its body diameter should be 33103096 m (2.6 times that of the earth), giving it a gas-cloud cover thickness of 8070452 m

Using this information, the body of Uranus is spinning at 3.7918E-05 radians per second and its properties are:

Δ = 0.28   ρ = 5392 kg/m³

8.9.12 Pluto

Δ = 8.64241984998   **ρ = 1859.960193 kg/m³**

Pluto is a small ice planet in which all of its matter exists in solid form, hence its relatively high density.

Having acquired a substantial sub-satellite population, the largest of which (Charon) is more than 10% as massive as the planet itself, the competing kinetic and potential energies are pulling Pluto into a localized orbit, and is the reason it has a 'Δ' value greater than 1.

Due to its localised orbit, the sun's potential energy is not acting at the planet's centre. I.e. there is no conflicting core-mantle spin to generate internal heat (energy), otherwise, Pluto would actually be a gas-planet.

Pluto is more entitled to be called a planet than either Mercury or Venus, because whilst all three are solid lumps of matter, at least Pluto has managed to acquire some moons; at least five in-fact. Moreover, if it had attracted less massive sub-satellites, it would have an active core, just as all the other planets in our solar system; except Mercury and Venus.

Whilst there are planets outside Pluto's orbit; MakeMake, Haumea, Eris, etc., as we know very little about them, I have not addressed them here.

8.9.13 Our Moon

Δ = 0.554903434   **ρ = 3343.599878 kg/m³**

As our moon comprises very different matter (3343.6 kg/m³) from the iron planets and its 'Δ' value shows that it is unlikely to be hollow, it must have come from somewhere else in the galaxy or solar system and therefore must have been acquired after the earth began to orbit its sun.

However, given the estimated density for the largest mass in the asteroid belt (Ceres) is 2090 kg/m³, it is a fairly safe bet that our moon originated from outside our solar system, especially given the inclination of its orbital plane.

8.9.14 Phobos

Δ = 0.275895222   **ρ = 1827.4186 kg/m³**

The same would normally go for Phobos, but its density is very similar to that of Ceres, so Phobos may be left over from the impact that created the asteroid belt, especially given their orbital proximity.

8.9.15 Deimos

Δ = 0.01434654   **ρ = 1471 kg/m³**

Deimos is probably hollow and also appears to comprise matter different to that which would be expected in its part of the solar system. But if the Asteroid planet was capable of creating rock, it is possible that Deimos was part of the Asteroid's upper mantle material, or perhaps one of its moons.

The Universe

9 Fact & Fiction

There are a number of myths prevalent in the scientific community, not least the Theory of Relativity and Quantum Theory. These obscure theories have ultimately led to mystical features such as anti-matter, sub-atomic particles, dark matter, Black Holes (which are actually black bodies) and singularities.

9.1 Sub-Atomic Particles (fiction)

These sub-atomic particles along with electron uniqueness; s, p, d, f, l, m, spin, etc. were invented to hold Bohr's atom together and make it work. Whilst sub-atomic particles, such as gluons, fermions, bosons, positrons, etc. may or may not exist, none of them are required to make the universe (as we know it) work; they are *unnecessary*. And if there is one thing certain about nature, it doesn't waste energy on things it doesn't need.

The real atom doesn't need them, energy transfer (electro-magnetic radiation and fields) doesn't need them, galactic structural integrity doesn't need them, so why would they exist?

For example; there is no need to split a proton into quarks (1-Down and 2-Up). An electrical charge will *always* have a magnetic counterpart (a magnetic charge); it is nature's fundamental design. It is a perfectly understandable, workable and reliable particle; it possesses everything it needs. There is nothing to go wrong and it is 100% efficient.
Why would you make it complicated and therefore unreliable by giving it component parts it doesn't need, that could go wrong and wastes energy?

Not only are these particles unnecessary, there is no physical evidence to show they exist. E.g. *Bubble-Chamber* diagrams reveal interactions that can be easily explained as collisions between electrons in free-flight.

The Hadron Collider (CERN) was built, at enormous cost, to identify and define all these sub-atomic particles, which is a shame because they don't exist, or at least they are not necessary to make the real atom work.

Anti-matter is also a fallacy. There is absolutely no argument that can explain or justify its need in our universal system.

The real atom works perfectly because it is a brilliantly elegant, simple, reliable orbital system of just two *perfect* particles; the electron and the proton. Everything we see and feel around us can be explained by them.

The atom described by Quantum Theory, which is not based upon orbital systems, cannot work because it cannot transfer energy.

Together with the *uncertainty* principle, the invention of these particles was needed to justify an atomic system that cannot work without them. Moreover, they remain unproven and undiscovered to this day.

9.2 Black Holes (fiction)

There is no such thing as *'Black-Holes'*, singularities or event horizons.

There are only *Black-Bodies*, which are celestial bodies that have no force-centre or significant satellite population. Black-Bodies can be any size because their inability to emit light is not based upon their ability to trap *'photons'* (light is radiated electro-magnetic energy and photons do not exist). But; galactic force-centres do generate sufficient core pressure to fuse small atoms together, but fusion is not a source of heat.

Stars do not have the core pressure to generate fusion.

The reason stars emit intense heat and light (electro-magnetic energy), is because stars are in orbit about a force-centre with a substantial satellite population of their own. The heat energy generated by the resultant internal friction unites protons and electrons as neutrons, creating neutron-rich atoms and ultimately fission; this is the source of our solar heat and light.

Fission releases the electro-magnetic energy we see and feel as light and heat, creating hydrogen and neutrons in the process. As more and more of a star's matter is converted to hydrogen, it will grow in *physical* size (not in mass). When there is insufficient viscous matter to generate the frictional heat at its core the star will become inactive (cold). This *black-body* will eventually become a hydrogen (proton-electron pair) cloud surrounding a neutron-rich core.

Galactic force-centres, which are by definition *Black-Bodies*, spin very slowly indeed. Hades, for example, is spinning at about $2E-07$ radians per second.

9.3 Big-Bang (fact)

The energy released within the ultimate-body has popularly become known as the 'Big-Bang'.

The 'Big-Bang' originated from the energy released by compromising the innermost core neutrons within an ultimate-body of >2.80364E+75 Quanta. The core of this body is where all the heaviest universal elements are created. It is only at the core of this mass that there is sufficient pressure to generate the compressive energy needed for heavy-element fusion.

The pressure at its core was sufficient to overcome the coupling ratio (φ) splitting the innermost neutrons into their component parts and releasing their stored energy in the form of alpha particles. The resulting chain reaction released >7.3544E+60 Joules of energy that caused the ultimate-body to explode into the lump-masses that became the celestial bodies we see today, and which in turn, are composed of the elements and molecules that were created during the previous universal period(s).

The largest ejected masses became [galactic] force-centres, attracting neighbouring smaller bodies to their vicinity through potential energy and creating the galaxies. The kinetic energies in bodies of varying mass caused each to orbit at different velocities and therefore at different orbital radii (Newton) and became galactic satellites.

Galactic satellites become galactic comets when destroyed through impact via orbital precession. These galactic comets are trapped by other galactic satellites and become planetary or lunar satellites.

The relative galactic velocities tell us that they are all moving away from each other, reflecting the ellipsoid (3D) nature of the universe following the original 'Big Bang'. This movement is the basis of Hubble's law.

Post 'Big-Bang', magnetism (gravity) is gradually retarding velocities and eventually all universal matter will re-collect back to the ultimate-body; and 'bang', the process starts all over again. This is a never-ending cycle that can continue eternally with no artificial (outside) help, and it complies with all three laws of thermodynamics.

The Universe

The 'Big Bang' occurs when Newton's attraction force $(G.m_p^2/R^2)$ exceeds Coulomb's repulsion force $(k.e^2/R^2)$
Where:
G is Newton's gravitational constant
k is Coulomb's constant
e is the elementary charge
m_p is the mass of a proton
R represents the *diameter* of a proton (two adjacent radii)

Together, these formulas define the *mass* necessary to compromise two adjacent core-neutrons:
$m_u > k.e^2 / G.m_n.\varphi + m_n > 4.68687882273808E+48$ kg
Where:
m_u is the *mass* of the ultimate body
this gives us 2.80059013353655E+75 proton-electron pairs (or neutrons)
if the average neutronic ratio is ≈1.147962 (iron), there are ≈1.496754E+75 neutrons in the universe

The neutron energy contained within the ultimate body and therefore dispersed into the universe during a 'Big-Bang' event, is;
$E_u \leq E_n . N / (2-\psi_{ave}) \leq 2.45146787204436E+62$ J
which was released as kinetic energy of varying magnitude in all the dispersed matter.

Assuming only 3% of neutron destruction (Little-Boy) the explosive energy of the 'Big-Bang' is likely to have been:

$E_u \leq e.N_p \leq 7.35440361613308E+60$ J

If the *mass* of the ultimate-body prior to the explosion is the same as the mass in the universe today (equivalent to 8.784256E+10 Milky Way galactic masses) the average velocity of all galaxies must be equal to $\sqrt{[2.E/m]}$ relative to the centre of the explosion, i.e.:

$v \leq \sqrt{[2.E_u/m_u]} \leq 1773498.4104391$ m/s

9.4 Dark Matter (fiction)

Almost all of the world's physicists believe that 85% of universal mass is *dark matter*, and that this dark matter comprises sub-atomic particles that cannot be seen, which is the reason it is called *dark*. This has arisen because early in the 20th century, a couple of these physicists (Fritz Zwicky & Jacobus Kapteyn) claimed that Newton's laws of orbital motion predicted that the Milky Way's stars should be thrown into outer space because of centrifugal force based upon its observed star-system population.

The large force-centre at the core of the Milky Way galaxy (Hades) was unknown at that time and was no doubt omitted from these physicists' calculations. However, they should have postulated that there *must* be a mass at the centre of *every* spiral galaxy; because that is how orbits work. I suspect, therefore, that the reason dark matter was invented, is simply that these scientists didn't understand Newton's laws of orbital motion very well.

Given that according to these laws, *only* force-centre mass defines orbital shape, and star-system population *only* defines a force-centre's spin-rate; I find the whole concept of dark matter very difficult to accept because galactic force-centres were unknown 100 years ago and planetary spin theory has only just been resolved.

Moreover, as we now know that electrons do not emit light, we cannot assume that 'black bodies' must necessarily be large enough to trap electrons, they are simply cold enough not to emit electro-magnetic radiation (e.g. light) and could therefore be *any size*.

A galaxy could therefore have a higher star-system population than observed as it may contain cold bodies that cannot be detected. This does not alter the fact, however, that Newton's laws of orbital motion and spin theory *never* predict that galactic systems will fly apart as a result of centrifugal force.

So; why do so many people still believe in dark matter when we should understand these laws much better by now?

The reason that none of this dark matter (sub-atomic particles) has yet been described or discovered is because it doesn't exist.

9.5 The Birth of Our Solar System (fiction)

It is currently claimed that our solar system was born from a cloud of gasses that a *'galactic-force'* pushed together into our sun (and planets). The sun then began to spin of its own accord and drag the planets around with it.

It is also claimed that gravity accreted our earth from rocks flying about in the solar system, and that this accretion process generated the heat that currently resides within the planet and that this hot planet slowly became covered with a shallow sea as it cooled down.

There are so many problems with this model it is difficult to know where to begin, but I'll have a go with a few of the obvious ones:

1) What is generally referred to as natural hydrogen (H^+), but which is simply a lone proton, is the only atom (or molecule) that can exist as a gas at ≈2.7255K (in open space). And accretion from lone proton is impossible (see 3) below).

2) A *'galactic-force'* cannot be transmitted through a vacuum.

3) Lone protons (H^+), which constitute 99.7% of what is today referred to as hydrogen, cannot accrete irrespective of how hard you push them together. Each proton is applying 2.27E+39 (1/φ) times more electrical charge repulsion than magnetic charge attraction can apply.

4) Rocks are only created within planets that have sufficient satellite mass to generate the internal frictional heat and consequent volcanic activity. So where did earth's rocky meteorites come from?

5) Without a moon, the earth would have had no internal heat to create volcanoes and earthquakes and hence no atmosphere.

6) More than 99.96% of earth's atmosphere has been generated by its surface life (over the last billion years) and potassium breakdown (over its lifetime). The remainder was created by volcanism which has only been possible since it has acquired a moon, which was long after its *'ejection'*.

7) Impact alone does not generate heat, especially in a planet with no atmosphere.

8) Because the early earth existed without a moon, it must have been cold.

The Universe

9) Spin theory is the only viable answer to the gas planets that can be demonstrated mathematically. So, the earth's internal heat can only be generated by its moon and is the reason Venus and Mercury are the only planets in our solar system that have no magnetic field.

10) The first law of thermodynamics tells us that energy cannot be created from nothing. Therefore, the sun cannot spin without an energy causing it to do so. If, as is currently claimed, the sun is dragging its planets around with it, this law also tells us that it should be spinning slower than can be attributed to Hades alone. However, it is actually spinning much faster.

11) If 9) above is correct, the same argument must apply to our sun (and all the stars).

12) Fusion does not generate energy, it requires energy input to work. So, it cannot be possible that our sun evolved from hydrogen (H^+).

13) If stars (including our sun) were created from natural hydrogen (H^+), how can we see their surfaces? Lone protons cannot collect, generate or emit electro-magnetic energy. Electro-magnetic energy can only be generated from proton-electron pairs (H), which are not natural hydrogen atoms.

14) Proton-electron pairs of hydrogen gas (H) are created by fission – they are the last remaining proton-electron pair after converting and removing all the other proton-electron pairs from an atom (as neutrons).

15) The only difference between stars and planets is that stars are brighter due to the orbital systems that generate greater internal heat.

I could continue but it is unnecessary. The current generally accepted model for the universe *must be* incorrect.

Stars are not *created from* hydrogen (H^+) through fusion, they are actually *creating* hydrogen (H) through fission.

All universal bodies originally comprised the same matter, that of the ultimate-body. They look and behave according to their orbital systems.

Our solar system is not 4.6bn years old; it is as old as the current universal period (*perhaps* 13.6bn years)

10 Primary Constants

There are very few primary constants, i.e. those that we must take for granted and on which *all* others are based; these are listed below

Symbol	Value	Units
e	1.60217648753E-19	C
electrical charge (elementary charge unit)		
m_e	9.1093897E-31	kg
magnetic charge (the mass of an electron)		
R_n	2.81793795383896E-15	m
distance (the neutronic radius)		
t_n	5.90596121302193E-23	s
time (neutronic period)		
ξ_m	1836.15115053207	
static ratio {m_p/m_e}		
ξ_v	1722.0458764934	
dynamic ratio {c/v_o}		
Σ	3E-91 (exact)	m^6
particle constant		

Table 10.1

All other physical variables and constants can be calculated from them.

Note: Temperature is not a real property, it is an interpretive value for the kinetic energy of the electron(s) orbiting in an atom's innermost shell.

10.1 Electrical Charge (e)

The static electrical charge held by all electrons and protons is the source of all universal electrical energy.

The fundamental unit; 'e' is held by all electrons and lone protons (H^+).

Its magnitude was discovered by Charles-Augustin de Coulomb.

After a lone proton has acquired an electron partner (H) the electrical charge generated by the partnership (e') will increase with increasing electro-magnetic energy (temperature; $\underline{T}$) thus:

$e' = m_p.RC \cdot \sqrt{[\underline{T}/\underline{T}_n]} = m_p.RC \cdot v/c$

10.2 Magnetic Charge (*kg*)

There is no such thing as mass,
i.e. **mass** is magnetic charge, the magnitude of which is equal to the elementary charge unit (e):
$m = |e|$
$m' = |e'|$

Therefore; what is currently referred to as the elementary charge unit should be the *electrical charge unit* (±e), and the unit of mass should be the *magnetic charge unit* (*m*). The magnetic charge's non-polar nature is what causes all particles to attract all other particles.

Every particle holds a constant non-polar magnetic charge and also *retains the capacity* to hold the same magnitude of electrical charge.

For example;
The electron holds '*m*' Coulombs of magnetic charge and '-e' Coulomb's of electrical charge constantly.
The proton holds '*m*"' Coulombs of magnetic charge and '+e' Coulomb's of electrical charge. However, its greater magnetic charge, gives it the capacity to increase its electrical charge to '+e" if and when partnered by an orbiting electron.

The number of particle assemblies (electron + proton + neutron) in a body:
$n = mass / 2 \cdot (m_e + m_p)$

The magnetic charge in each particle assembly:
$m = 2 \cdot (e + e_n) = 5.88688075484235\text{E-}16$ C

Using the planet Mercury to demonstrate this concept ...

First; we generate a new magnetic constant (M) based upon Coulomb's constant (k) as follows:

$M = c^2 \cdot m_e^2 \cdot R_n / m_p \cdot e^2 = 4.89477777726655\text{E+}06$ kg.m³ / C².s²

Note:
If $R_K = \sqrt{[G/M]} = 3.69243914581423\text{E-}09$ C/kg
{√[(m³ / s².kg)/ (kg.m³ / C².s²)] = C/kg}
and RC = 1.75881869180547E+11 C/kg
$R_K / RC = 2.09938589066497\text{E-}20 = \sqrt{\varphi}$
i.e. the square-root of the coupling ratio

Next: we calculate the magnetic charge in Mercury and its force-centre:

	Mercury	**Sun**
mass (kg)	3.3011E+23	1.9885E+30
No. of particle groups	9.862670275E+49	5.941025671E+56
magnetic charge (C):	m_2 = 1.2189111E+15	m_1 = 7.3424152E+21

Table 10-2: Magnetic charge of the planet Mercury and its force-centre

Finally: we use our new constant 'M' and the magnetic charges (m_1 & m_2) to calculate the 'gravitational' force between Mercury and the sun:

$F = M.m_1.m_2/R^2$ (R is the separation distance at Mercury's perigee)

F = **2.07016816968015E+22 N**
(F^p = **-2.0701682E+22** $N^{\#}$)

and it works for all the planets in our solar system

There is an additional argument:

Potential Energy:
The potential energy of every orbiting body may be calculated using the following formula:
PE = [($m_1.m_2$) ÷ (m_1+m_2)] . g.R {J}   **#1**
Where; g is the gravitational acceleration from m_1 at radius R
This formula (**#1**) produces exactly the same result as Isaac Newton's formula for the same property:
PE = $G.m_1.m_2/R$ {J}

PE = √ [($E_1.E_2$) ÷ (E_1+E_2)] . $g.R/v_c^2$ {J}   **#2**
E_1, E_2 & v_c are the energies of a star and its satellite and v_c is a constant[#]

Both formulas (**#1** & **#2**) produce exactly the same potential energy between every force-centre and every satellite in our solar system confirming that mass may also be defined as magnetic energy.

Moreover, if 'n' represents the number of atomic particles in a celestial body and e is their Coulomb value; for every planet in the solar system:
$g.R.e.(n_1.n_2)/(n_1+n_2)$ = 2.08908021007445E-08 {**C.m²/s²**}
Magnetic Energy!

[#] *refer to Appendix; Philosophiæ Naturalis Principia Mathematica Rev. IV*

10.3 Distance (R_n)

The neutronic radius, which is achieved by an orbiting electron when travelling at 'c', can *only* be explained using Newton's laws of orbital motion and Coulomb's law of electrical force. It occurs in far too many constants (magnetic, permittivity, Rydberg's, Planck's, Coulomb's, Henry's, etc.) to be rejected as a *fundamental physical constant.*

The neutronic radius is also the basis of $E=mc^2$

The conversion of mass to energy with velocity together with the space-time/gravitational distortion around force-centres as defined in Relativity, would render such an orbital radius impossible. I.e. the electron would be orbiting inside the proton at 'c' and R_n would be incorrect, making magnetic constant, permittivity, Rydberg, Planck, Coulomb, Henry, etc. incorrect, which we know is not the case.

10.4 Time (t_n)

The fundamental unit of time is the period required for an electron to complete one full orbit of its proton partner at the '*speed of light*' (c).

10.5 Static Ratio (§m)

The static ratio defines the relationship between the mass of a proton and the mass of an electron, but because both have the same density, it also describes their relative volumes:

$§m = mp/me = Vp/Ve = 1836.15115053207$ no units

This value is very specific. It defines the orbital instant when magnetic field energy generated in the proton-electron pair equals the centrifugal energy in the electron. Which occurs when the electron is orbiting at the velocity of electro-magnetic energy (c) and simultaneously achieving the neutronic radius (Rn).

If §m > the above value; the orbiting electron would impact its proton partner before it achieved 'c'

If §m < the above value; the orbiting electron's velocity would achieve 'c' before it achieved Rn and thereafter remain constant

In both cases, neutrons could not be created and our universe would not exist.

10.6 Dynamic Ratio (ξ_v)

The dynamic ratio defines the relationship between 'light-speed' velocity (c) and Planck's minimum velocity (vo):

$\xi v = c/vo = 1722.0458764934$ no units

It was found from:

Planck's constant; $h = \sqrt{(\pi.m_e.e^2.a_o / \varepsilon_0)} = 6.62607174469163\text{E-}34$ kg.m²/s

and;

Rydberg's constant; $a_o = \varepsilon_0.(h/e)^2 / \pi.m_e = 5.2917721067\text{E-}11$ m

But is derived from:

$\xi_v = 4\pi.\sqrt{[ao/Rn]} = 1722.0458764934$

ξ_v is the factor that defines Planck's key orbital radii of an electron:

Note:

$h = \sqrt{[\pi.m_e.e^2.a_o / \varepsilon_0]}$ {J.s incorrect units}

to:

$h' = \frac{1}{2}.m_e.c^2.R_n$ {J.m}

Cold: $R_c = R_n.\xi_v^3$

Planck Maximum: $R_o = R_n.\xi_v^2$

Planck Mean: $R_m = R_n.\xi_v$

Planck Minimum: R_n

Rydberg's Radius: $a_o = R_n.\sqrt{[\xi_v/4\pi]}$

10.7 Universal Constant (Σ)

$\epsilon_e = m_e/m_u$ {no units} or $\quad \epsilon_e = m_e/\rho_u$ {m³}

$\epsilon_p = m_p/m_u$ {no units} or $\quad \epsilon_p = m_p/\rho_u$ {m³}

$\xi_m = \epsilon_e/\epsilon_p$

$\Sigma = 1/(\epsilon_e \cdot \epsilon_p)$ {no units} or $\quad \Sigma = (m_e \cdot m_p)/\rho_u^2$ {m⁶}

In reality, this universal constant is a little bizarre, in that ϵ_e & ϵ_p are the numbers of electrons or the number of protons in a unit mass of ultimate density respectively.
But it also seems to tie Newton's and Planck's particles together.

It has such a bizarrely accurate value, however, that it may or may not exist. The trouble is, it seems to appear everywhere. It even gives us 'G':

$G = \sqrt{[\Sigma \cdot a_o^2 \cdot c^4 / m_p \cdot m_e]}$
$m_e \cdot m_p = \Sigma \cdot \rho_u^2$
$\rho_u = m_e \cdot \sqrt{[\xi_m/\Sigma]}$
$V_e \cdot V_p = \Sigma$
$V^{P2} = r^{P6} \cdot (4/3 \cdot \pi)^2 = \Sigma$
$r^P = \sqrt[6]{[\Sigma / (4/3 \cdot \pi)^2]} = 5.07563837996471E-16$
$\quad$ is the radius of a Planck particle
$\xi_m = (\varphi \cdot R_n / a_o)^2 / \Sigma$
$\xi_m = \Sigma \cdot (\rho_u/e)^2$
$\xi_m = \varphi^2 \cdot (4\pi/\xi_v)^4 / \Sigma$
If $F^N = G \cdot m_e \cdot m_p / a_o^2$ then; $F^N/F^P = \Sigma$

Its units *appear* to be m⁶ (see above). I say "*appears*" because it also unites the Newton and Planck forces (F^N & F^P; see above), which means it can also have no units.

Σ also defines the ratio; "*electron orbital radius*" : "*radial separation (centres)*"; ϑ

That is; the relative radii of the orbiting electron and the "electron radius plus proton radius" in the neutronic condition:

$\vartheta = R_n / [r_e + r_p] = 1.46677550700175$

The Universe

For example, at a velocity of 'c':

Σ	3E-91	{m⁶}
ϑ	1.46677550700175	
G	6.67359232004334E-11	m³ / kg.s²
$ρ_u$	7.12660796350450E+16	kg/m³
φ	4.407421117923340E-40	
r_e	1.45046059426276E-16	m
r_p	1.77613270336827E-15	m

All of which are correct!

10.8 Temperature (T_n)

Temperature is not a genuine property. It is a convenient way of describing the kinetic energy in the electrons of an atom (in the innermost shell).

$$T_n = m_e.c^2 / k_B.\sqrt[3]{[½.\xi_v]}$$

Where:

'$m_e.c^2$' represents the potential energy between a proton and its orbiting electron at the time they unite to create a neutron

and;

'$k_B.\sqrt[3]{[½.\xi_v]}$' represents the factor to define temperature (k_B is Boltzmann's constant)

The Universe

11 Model Verification

The following sub-chapters provide supporting verification of the atomic model described in this publication.

11.1 Density vs Temperature

The magnetic field energy (MFE) generated by proton-electron pairs holds adjacent atoms together (viscosity) and is constant irrespective of temperature. The electrical charges (EC) generated in protons repel adjacent atoms (gases) and varies between e and e' with temperature.

Given that temperature effects (e.g. gasification) are dependent upon repulsive EC between adjacent atoms and density is dependent upon an attractive MFE between those same atoms, and that both charges are created by the same energy generation process (proton-electron pairs), density and temperature should follow similar patterns of behaviour according to the number of nucleic protons (atomic number (Z)).

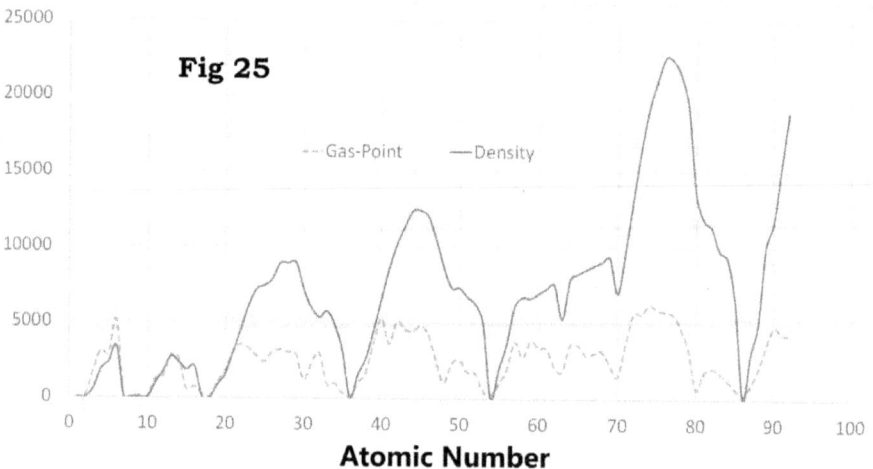

This relationship (between EC and MFE), which can clearly be seen in the temperature/density vs atomic number plot shown in Fig 25 is governed by the nucleic structure, which is the last significant piece of the atomic puzzle.

Moreover, it can also be seen that atomic attraction due to MFE is indeed constant whilst the EC is variable, based upon atomic size (proton population) and temperature.

11.2 Specific Heat Capacity

The specific heat capacity of an atom defines the amount of energy it can hold in relation to its mass per unit temperature. This means the sum of the kinetic energy of all electrons in an atom's shells relative to its mass and 'temperature'.

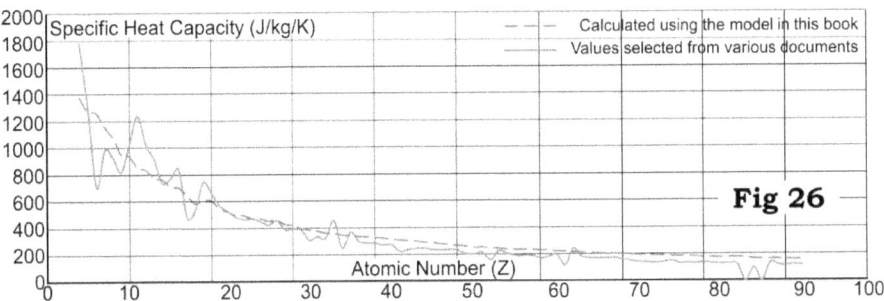

Fig 26

Fig 26 shows the calculated values for specific heat for all atoms from Z=4 to 92 compared to the documented values that have been taken from various sources and which are subject to experimental error.

This calculation technique, is as follows:

$$SHC = KE_T / Y.m.T_1 \quad \{J/kg/K\}$$

where:
KE_T = the total kinetic energy in every electron in the elemental shells
T_1 = the temperature of the electron(s) in the innermost shell
Y = temperature constant
m = the total mass of the atom (including electrons and neutrons)

11.3 Gas-Point

The gas-point of any atom is the temperature at which its electrical charge (EC) exceeds its magnetic field energy (MFE).

If the MFE is greater than the total exposed EC, the atoms will exist as viscous matter; otherwise they will exist as a gas.

Outlying nucleic neutrons protect adjacent atoms from EC. The more outlying neutrons, the greater the protection (higher gas-point temperatures and greater densities).

Density rises with atomic number (Fig 24) because larger atoms tend to collect a greater percentage of neutrons, due to the higher collective MFE within atoms.

Gas-point temperatures also rise with atomic number (Fig 27) for the same reason, but this rise is much less marked because only the outlying and exposed proton EC actively repels.

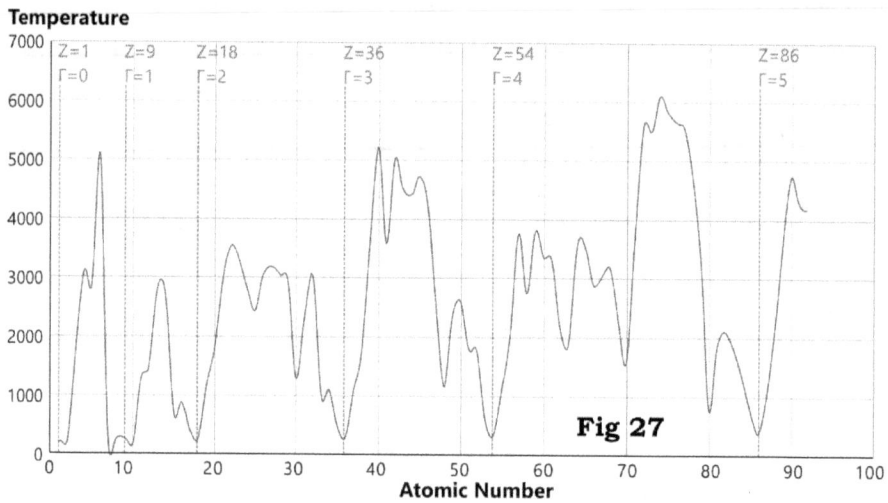

Fig 27

A mathematical relationship that reflects reality has been identified for all the atoms in the Periodic Table: $\Gamma = 9 \cdot (\psi - 1)$
Refer to Chapter 4.3 for as definition of 'ψ'

This relationship is the basis for mathematical chemistry.

The Universe

11.4 Our Sun

Almost all natural hydrogen exists as lone protons (H⁺), which cannot be solidified or liquefied due to their similar positive charges. Lone protons also have no way of absorbing or emitting electro-magnetic energy (heat and light).

It is currently claimed that our sun was created from a cloud of hydrogen atoms that accreted into a star due to potential energy and an *external force*. This is of course, impossible.

The surface temperature of our sun is said to be about 5778K, which would be impossible if the sun's surface comprised lone protons that cannot collect or emit electro-magnetic energy (i.e. heat or colour).

Yet according to the atomic model proposed in this book, at 5778K;
KE = 3.7493802154296E-19 J
electron velocity = 912757.252 m/s
at an orbital radius of 3.03992067E-10 m
$f = v / 2\pi R$ = 4.77873747733E+14 Hz
$\lambda = c/f$ = **6.2734657516211E-07 m**

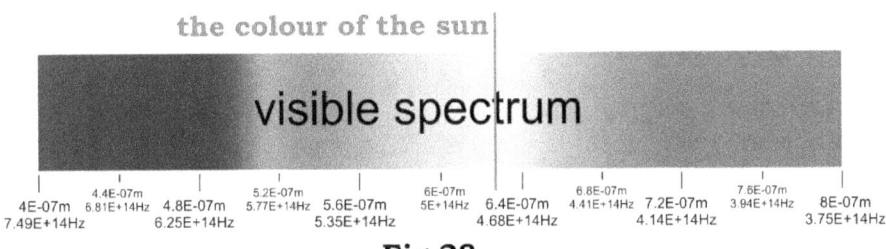

Fig 28

This demonstrates that the hydrogen at the surface of the sun is predominantly proton-electron pairs, which in turn means that these hydrogen atoms (proton-electron pairs) *must be a by-product of fission.*

11.5 PVRT

The final and most compelling proof of this atomic model can be seen by the replacement of the well-known theory; 'P.V = n.R_i.T' with an alternative calculation method using the potential energy in a proton-electron pair:

Coulomb's force between the electrical charges in adjacent protons is responsible for generating gas pressure.

Dalton's law states that each gas in a mixture of different gases will evenly fill its container independently of all the other gases in the mixture.

Partial pressure theory states that the total pressure of a gas mixture is the sum of the pressures of each individual gas.

Both of these statements appear to conflict. However, because *all* adjacent atoms are subject to electrical repulsion (Coulomb's law), but *only atoms with the same nucleic structure* (refer to Chapter 4.1) will be subject to a balanced pressure, both of the above statements can be rationalised.

Today, we calculate the pressure (p) of a gas thus:
p = n.R_i.T / V
where: 'n' is the number of moles in the gas, 'R_i' is the ideal gas constant, 'T' is its temperature and 'V' its volume.

But it can also be calculated thus:
p = ρ.PE_1 / m_M.Y = T.m_e.ρ / X.Y.m_M = k_B.T.ρ / m_M

where: 'ρ' is the gas density, 'PE_1' is the potential energy between the proton and the electron in shell-1 and 'm_M' is the molecular mass:
PE = m_e.v_e^2
v_e = $\sqrt{[T/X]}$
X & Y are heat constants

which provides *exactly* the same result as the PVRT calculation method but is much simpler because there is no need to play with moles.

Because this latter calculation method replaces 'PVRT' altogether, along with the need for Boltzmann's constant, Avogadro's number, gas temperature and the ideal gas constant with potential energy, the model described in this book must be considered correct.

12   The Laws of Thermodynamics

The First Law of Thermodynamics: Conservation of energy
Energy can never be lost, it can only be transformed or transferred.

The Second Law of Thermodynamics: Heat will not spontaneously pass from a colder body to a hotter body

A high-energy source (hotter body) will spontaneously lose energy to a low-energy source (colder body) but you must add work if you want energy to transfer in the other direction (up-hill so to speak). This law essentially states that it is impossible to create energy from nothing.

This law also claims that energy can, and in fact is, lost by a system to its surroundings but that the reverse cannot happen i.e. an increase in disorder is an inevitable feature of time.

The Third Law of Thermodynamics: The entropy of a substance approaches zero as its temperature approaches zero (absolute)

Entropy is the term used to define disorder. The higher a substance's temperature the more disordered will be its atomic structure and the higher its entropy. E.g. gas has a higher entropy than a solid substance.

The Universe

The Universe

13 So; What Now?

Given what we now know about universal energy;
1) How it is created (orbits and spin-friction)
2) Where it is created (stars and planets)
3) How it is transmitted (electro-magnetic energy)
4) Where it is stored (neutrons)

We now have access to unlimited, clean, free energy sources;
1) Elliptical Orbits
2) Mantle heat
3) Neutrons

Moreover, these theories can give us the ability to *mathematically* predict chemical reactions in *all* matter irrespective of complexity; the *molecular calculator*.

Such a calculator would preclude the need for material, chemical or pharmaceutical testing and experimentation. No more risk, material, time or money need be wasted on such activities and every country in the world would be able to design [100% accurate] new materials, chemicals and medicines in safety, from a computer terminal with trained but semi-skilled personnel. Furthermore, the creation of comprehensive organic and inorganic chemical databases will remove the need for duplicate effort together with the horrendous qualification periods for new medicines imposed by various national and international health authorities.

Because we now know where the universe stores its energy, we have access to an unlimited supply free from waste and pollution. We could do something useful with the world's nuclear waste, such as fuel for clean, controllable, efficient energy generators of any size. Much less mining!

Moreover, due to the discovery of the true meaning of $E=mc^2$, there is no longer any reason to assume that light-speed is a limiting condition for matter. And if matter has no mass, imposing a limiting velocity owing to the conversion of mass to energy becomes unnecessary. The speed of light is simply a speed for electro-magnetic radiation, such as that for sound: there's no reason it cannot be exceeded.

Anti-gravity also becomes *theoretically* possible. All you need to do is repel the earth's *magnetism*, which is easier than opposing *gravity* with mass.

The Universe

A few of the possibilities from the discoveries explained in this book are listed below?

1) Molecular calculator (and database) giving new (perfect) materials, medicines and chemicals in minutes
2) Limitless, clean, free, safe energy (by-product = hydrogen)
3) Propulsion-free satellites
4) The ability to safely recycle nuclear waste
5) Energy cells that can be fuelled with any matter (e.g. rocks!)
6) Alter elements into something else
7) Change the colour of matter electrically
8) Together with PERS#, the elimination of skin-friction offers virtually free travel
9) Perfect lubricants (machines with almost eternal life)
10) Free energy from the earth's mantle
11) Massive reductions in: pollution, material waste, energy, etc.
12) Travel by impulse drive, replacing all other forms of transport & transit.

PERS = *potential energy recovery system*

In other words, we now have the ability to ...
... massively reduce energy and battery production;
... massively reduce mining requirements;
... massively reduce transport costs;
... massively reduce the number of chemical laboratories;
... eliminate; national power stations & transmission lines, wind-turbines & solar panels;
... eliminate pollution from energy generation;
... create vehicles with no engine or drivetrain that need no refuelling;
... create 100% recyclable packaging

All the energy we use today requires the generation of much more to harness and recycle it. Instead of generating energy at an efficiency of less than 10%, we now have access to energy generation that is 231,000,000% efficient.

Instead of swapping one pollutant for another and/or simply moving it around as we do today, we could create a genuinely clean place for everyone in which to live; together with limitless cheap energy for all.

Appendix

References, symbols, glossary, etc. used throughout this book.

The Universe

A1 References

Apart from the work carried out by the (pre-20th Century) scientists listed in Appendix A3, all the theories postulated here are new; this book constitutes a rewrite of the natural laws of physics.

References to the *Heroes'* works are too numerous to mention here, but they are all universally known and accepted. Except for Max Planck's work, the problems with today's physics is due entirely to the theories espoused since the beginning of the 20th century, most of which are flawed.

Philosophiæ Naturalis Principia Mathematica Rev. IV; Keith Dixon-Roche; 978-1-07215-605-5

The Atom; Keith Dixon-Roche; 978-1-08610-029-7

The Neutron; Keith Dixon-Roche; 978-1-08251-683-2

The Physical Constants; Keith Dixon-Roche; 978-1-79422-609-8

The Life & Times of the Neutron; Keith Dixon-Roche; 978-1-08239-479-9

The Universe; Keith Dixon-Roche; 978-1-70753-878-2

but some additional sources are listed below:

Magnificent Principia; Colin Pask; 978-1-61614-745-7

Seven Brief Lessons on Physics; Carlo Rovelli; 978-0-141-98172-7

Science Data Book; Open University; 0 05 002487 6

Science and Technology Dictionary; Chambers; 0-550-18026-5

A Dictionary of Scientific Units; H G Jerrard & D B McNeill; 0-412-28100-7

Some additional sources are listed below:

Magnificent Principia; Colin Pask; 978-1-61614-745-7

Science Data Book; Open University; 0 05 002487 6

Science and Technology Dictionary; Chambers; 0-550-18026-5

A Dictionary of Scientific Units; H G Jerrard & D B McNeill; 0-412-28100-7

The Universe

A2 Glossary

Atomic Particle	One of the three components that comprise an atom
Big-Bang	The eruption that occurred when the Ultimate-Body accumulated sufficient 'mass' to compromise the integrity of the innermost neutron.
Black-Body	A collection of Quanta too cold to emit electro-magnetic radiation in the frequencies that would enable detection.
Coupling Ratio (φ)	The ratio of the coupling force due to a magnetic charge and the coupling force due to an electric charge: $\varphi = G.m_p.m_e \div k.e^2$
EME	Electro-magnetic energy (radiation)
Gas	Atoms that possess greater electrical charge energy than magnetic field energy
Great Attractor	The residual matter left over from the last 'Big-Bang'
Hades	The Milky Way's force-centre
Proton-electron pair	A proton that hosts an orbiting electron
Speed-of-light	The velocity of electro-magnetic energy (c = 299792459 m/s)
Sub-Atomic Particle	The many particles said to compromise atomic particles (leptons, gluons, fermions, quarks, etc.)
Ultimate-Body	A body that contains all the Quanta in the universe (≈2.8E+75) and represents the maximum single 'mass' that can exist without generating a Big-Bang.
Ultimate Density	The mass-density of all three atomic particles $\rho = 7.12660796350449E+16$ kg/m³ Nothing in nature has a 'mass'-density greater than this value
Universal Period	The time elapsed between Big-Bangs
Viscous	Solid or liquid matter in which magnetic field energy is greater than an electron's electrical charge

All other definitions can be found on the following web page:
http://calqlata.com/help_definitions.html

The Universe

A3   The Heroes

The heroes of this story, to which I offer my gratitude, are listed below

It is not necessary to identify the invaluable contributions made by each of these contributors, they are all widely known and available in almost every scientific publication in circulation today.

Nicolaus Copernicus (Polish) 1473-1543
William Gilbert (English) 1544-1603
Tyco Brahe (Danish) 1546-1601
Galileo Galilei (Italian) 1564-1642
Johannes Kepler (German) 1571-1630
Christiaan Huygens (Dutch) 1629-1695
Isaac Newton (English) 1642-1727
Edmund Halley (English) 1656-1741
Charles-Augustin de Coulomb (French) 1736-1806
Hans Christian Ørsted (Danish) 1777-1851
Michael Faraday (English) 1791-1867
Josef Stefan (Austria) 1815-1863
James Clerk Maxwell (Scottish) 1831-1879
William Crookes (English) 1832-1919
Ludwig Boltzmann (Austria) 1844-1906
Hendrik Lorentz (Dutch) 1853-1928
Jules Henri Poincaré (French) 1854-1912
Johannes Robert Rydberg (Swedish) 1854-1919
Max Karl Ernst Ludwig Planck (German) 1858-1947

The others that were instrumental in the completion of this book are:

My long-suffering wife (Brigitte) sub-editor and critic

My daughter (Eléonore), who initiated this project

Kenneth Pickering who first suggested that I write it

My thanks go out to all the above each of whom have provided a valuable piece of the puzzle without which the final solution would not have been possible, along with my sincere apologies to anybody I have unintentionally omitted.

www.ingramcontent.com/pod-product-compliance
Lightning Source LLC
Chambersburg PA
CBHW070555220526
45467CB00003B/1211